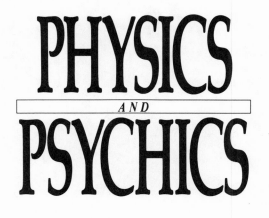

PHYSICS
AND
PSYCHICS

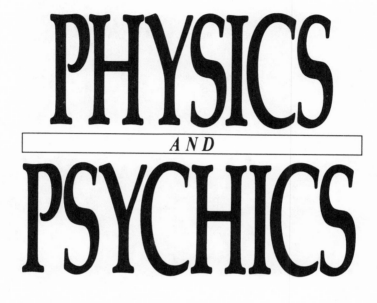

PHYSICS
AND
PSYCHICS

THE SEARCH FOR A WORLD
BEYOND THE SENSES

VICTOR J. STENGER

PROMETHEUS BOOKS
BUFFALO, NEW YORK

To the memory of Victor and Mary Stenger,
who never told me what to think.

Contents

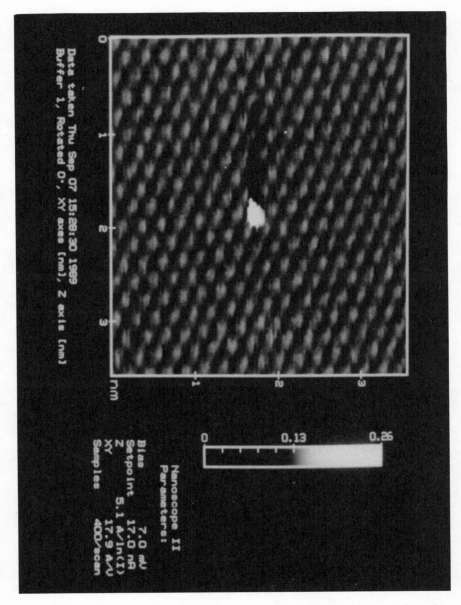

An Individual atom of the chemical element Chromium as photographed by a Scanning Tunneling Microscope. The area shown is 3 nanometers (3 billionths of a meter) square.
(Photo courtesy of Professor Klaus Sattler, University of Hawaii at Manoa.)

Preface

$$\Psi = 0$$

Like so many of the words we use to describe our deepest notions, "physics" and "psychic" are derived from the language of the ancient Greeks, who first gave rational expression to the concepts these words represent. Physics comes from *physis* (φυσισ), the Greek word for nature. Psychic is derived from *psyche* (ψυχη), the term for soul or spirit.

Physics refers to the objects of our senses, the world of body and matter. Phenomena thought to be associated with the nonmaterial world are called "psychic" (sometimes "psychical"), often abbreviated by the Greek letter psi (Ψ). The conclusion of this book can be summarized by a simple equation: $\Psi = 0$.

The Greeks were the first to apply the power of reason to the study of nature and soul. Their word *physiologia* translates as the science or logic of nature, although its modern derivative, "physiology," refers in more limited fashion to the study of the physical bodies of living organisms.

Similarly, the modern term "psychology" has become narrowed to the science or logic of mental phenomena, rather than the wider class of anything beyond the physical. Conventional psychology confines itself to the study of mind and human behavior independent of any assumption of a spiritual or supernatural component, although the prefix *psych,* with its ancient connection to soul or spirit, suggests otherwise. This unfortunate confusion results from the ancient belief in the equivalence of mind and soul. Since time immemorial, human behavior has been assumed to arise from mental and spiritual processes that transcend the world of matter and energy with which the physicist deals. But another possibility exists: Mental phenomena could be physical, wholly a property of the material world.

Mental processes are said to contain "inner" components of a different type than the "outer" components of the sensory world, and this view has led to a division of the studies of matter and mind. Nevertheless, matter and mind have come together in neurophysiology and the other neurosciences. In recent years these disciplines have begun to assume much of the burden previously borne exclusively by psychology, as the physical nature of mind has slowly but inexorably established itself.

Even more fundamental, classifying phenomena as physical and psychical parallels a distinction between two forms of reality that most people also take as self-evident, the division of the world into matter and spirit. The common belief is that two universes exist: (1) matter observed with our eyes and other senses, or scientific instruments, that seemingly obeys natural laws; and (2) another, invisible universe, undetectable by normal means, that is not bound by the laws of the material world—the transcendent universe of the spirit.

If physics is the study of matter, then we might term the study of a world beyond matter "paraphysics." The ancient term "metaphysics" could also be used, but it has other common meanings that suggest more than a mere extension of the methods of physics to phenomena beyond the realm of matter. Metaphysics can refer to any form of abstruse philosophy.

Paraphysics is a nice term, but I have never heard of anyone doing what I would regard as paraphysics: searching for the transcendent far out in space or deep inside atoms using the high-tech tools of the physics laboratory or astronomical observatory. Rather, most attempts to seek evidence for a world beyond matter center instead on human mental phenomena, as observed in the narrow range of everyday experience here on earth.

And so the term "parapsychology" has come to be used for those investigations that seek evidence for a nonphysical component of reality. Parapsychology, not conventional psychology, is normally associated with psi. In its purest interpretation, parapsychology might be termed "the science of the soul." Unfortunately, as we will see, parapsychology has not earned a reputation deserving of such a noble-sounding title.

Parapsychology has a built-in anthropocentric prejudice: that the particular kind of carbon-based life existing on our tiny speck of a planet has a special quality distinguishing it from ordinary matter. The assumption is made that human consciousness, or some extension of consciousness to other forms of life from cockroaches to plants, somehow interacts with the transcendent world of the spirit. I will call this assumption "the psi hypothesis."

That routes to a transcendent world may exist other than through consciousness seems not to have occurred to any parapsychologists whom I have read. But I can imagine such a world having nothing to do with living organisms. Perhaps that world connects instead to some yet undiscovered particles not residing within earthly biological materials, but existing in the centers of galaxies. Or perhaps transcendent connections will be found in some new force, still to be uncovered as future particle accelerators probe to higher energies and smaller distances. Of course, that's pure speculation, and hardly likely. My point is simply that the transcendent, if it exists at all, could have nothing to do with mental phenomena.

The psi hypothesis arises from the ancient belief that mind has powers that go beyond matter, that it is capable of performing miracles, that it interacts with other worlds through some special channel or sixth sense. Seeking scientific evidence for such powers is the goal of parapsychology. However, for parapsychology to be credible as the science of psi phenomena, it cannot take the existence of these phenomena as an established fact. Logic and honesty require parapsychologists to admit that convincing scientific evidence for the transcendent has not yet been found. Many, but not all, have admitted this.

Is there any basis for the belief in a world beyond the senses? Is there evidence for spirit or soul? Does the human mind possess supernatural or paranormal powers? Do mental processes—thinking, reasoning, dreaming—have aspects that cannot be understood in terms of our conventional scientific description of the material world? In other words, is there a nonmaterial world?

These questions have been explored over and over again throughout history. Support for the existence of a nonmaterial world can be found in the literature and customs of all ages and cultures. If we were to decide the issue by the weight of words on one side or the other, the scale would come down heavily on the side of the transcendent. For whatever reason, most human beings believe in a world beyond matter. But does that make it so? As Mark Twain noted, most people also once believed that the earth was flat.

Even in the modern age, influenced as it has been for centuries by scientific thought, when few still doubt that the earth is round, we find wide support for the idea of a transcendent reality. Newton believed in it, though he replaced the authority of Church and scripture with the authority of science. Einstein, like most scientists today, seemed to accept the notion of an underlying design to the universe, though his views were far from the traditional ones of his Jewish heritage.

Like many scientists today, Einstein often used the word "God" metaphorically in his writings, as another term for the order of nature. When an orthodox rabbi challenged the renowned physicist's beliefs, Einstein followed his wife's advice and answered that he believed in the "God of Spinoza." That got him out of trouble, even though Spinoza had been denounced for his belief that the divinity is nature itself.

Some scientists have lent their prestige and added their own words to the mass of those affirming supernatural beliefs—including some who claim to have discovered scientific evidence to back up their supernatural beliefs. We will meet the best known of them in these pages.

This book adds a few ounces to the other side of the scale, though I harbor no illusions that it will produce anything approaching balance. I write from the admittedly special (but I hope to convince you, not narrow) perspective of an elementary particle physicist. This perspective has been molded by my thirty-year participation in the giant research collaborations and massive experiments required today to roll back the curtain that conceals the fundamental nature of the universe.

Throughout my professional career, I have struggled to gain arcane knowledge, to convince granting agencies that my projects are worth funding, and to work and debate with dozens of collaborators with huge egos and differing points of view. I have had to convince my competitors and the tough referees of scientific journals that what my collaborators and I learned from the process merited acceptance into the ranks of scientific knowledge.

I have learned that obtaining new knowledge is not easy and that we shouldn't expect it to be easy. All the easy stuff was discovered a long time ago. Today, new knowledge is accumulated only through the greatest effort and concentration of resources, often by teams of hundreds of scientists.

From this perspective, I examine the evidence for the reality of anything beyond the universe revealed by our multimillion dollar scientific instruments. Those instruments currently yield a universe of quarks, leptons, and gauge bosons interacting with one another by means of the gravitational, electroweak, and strong nuclear forces.

Does this examination uncover any signal of a world beyond that of known matter—a signal capable of holding up under the same scrutiny applied to data gathered in all the conventional scientific disciplines where fundamental knowledge is at stake? Does anything that we have learned from physics, astronomy, chemistry, and the other physical sciences provide a confirmation of spiritual or psychic substances or forces, whatever their nature?

Does anything from the biological sciences indicate the existence of a "spark of life" or any special qualities of living beings that distinguish them from inanimate matter? Or is it possible that their high level of organization can be understood as a purely material phenomenon, developed by the natural, at least partially chance, processes of evolution and mechanisms of self-organization? Could this all have happened unguided by any power beyond the realm of matter?

From the behavioral and neurosciences, does anything we have learned about the mental processes of human beings or other living organisms force us to seek an explanation beyond that of atoms and electrons interacting inside the brain by means of well-known physical forces?

What about recent ideas that are still termed physical but postulate new holistic principles, or so-called "emergent properties," that operate on the global scale, controlling matter down to the microscopic level? Do they at least indicate that the universe is more than discrete atoms, that it is an unbroken whole with everything intimately connected to everything else?

And finally, does anything from the laboratory and field investigations of psychical research and parapsychology provide scientific evidence for an otherworld?

Now, many are likely to remark, "You can't prove or disprove that a world beyond matter exists." Strictly speaking, that is true, and I will not present here any such proof or disproof.

All anyone may reasonably expect is a careful, critical look at the evidence, and a rational judgment based on that evidence. To be credible, that analysis must ignore the insistent voices of those who want us to believe what they tell us to believe, simply because they tell us to. We cannot just take their word. We have to find out for ourselves.

I would be foolish to suggest that the current picture of the universe provided by the natural sciences is the final one. Surely it will continue to be modified, as each new generation of experiments using the latest technology uncovers further facts about nature. I fully anticipate that the future scientific view of matter will be substantially different from ours today.

Nevertheless, we have sufficiently progressed in our knowledge of the universe to be fairly confident that whatever direction the progress of science takes in the next century, we will not return to the irrational world of mystical and occult beliefs that currently exists as a remnant of the ignorant and superstitious childhood of humanity. If the transcendent is out there, it very likely will turn out to be something new, something never glimpsed in the deliria of history's mystics.

All things are possible, but many things are too unlikely, too preposterous, to take seriously. As a physicist, I can only believe what I see with my own two eyes, either naked or aided by the magnifying lenses of modern technology.

I have looked with my eyes at pictures taken with a scanning tunneling microscope and have seen atoms (see frontispiece), and so I believe in atoms and the fundamental discreteness of matter. I have looked at pictures taken with the most powerful telescopes showing distant galaxies, and so I believe in galaxies and the immensity of the universe. But nowhere in those pictures do I see the slightest direct or indirect hint of gods, spirits, or ghosts. These only seem to exist in the dreams and fantasies that rattle around in our heads.

Despite widespread belief to the contrary, no mystical revelation has ever told us anything about the universe that could not have been inside

the mystic's head all along. The most basic truths about the universe—
its size, constituents, the fundamental laws these constituents obey, and
humankind's place in it—are nowhere even hinted at in the sacred scriptures
that recorded the supposed revelations of history's leading religious and
mystical figures.

While belief in the special powers of the mind has been dominant
throughout history, it was by no means universal. The same ancient Greeks
who gave us science also first imagined a material universe subject to natural,
impersonal forces, with no reality beyond that universe. However, other,
more influential Greek thinkers continued to promote the idea of an invisible
reality that had, at that early time, been remembered from an even greater
antiquity.

The mystery of human thoughts and emotions largely accounts for
the longstanding belief in the duality of mind and body. Until the most
recent decades, people simply could not conceive of matter thinking. But
modern computer scientists have forced us to consider the very real possibility
that matter can think. Can the brain simply be an electrochemical machine?

Part of the reason for the common belief that mind can work miracles
is the huge number of people who claim they have seen such miracles
performed. Certainly these claims cannot be ignored. However, such testimony is anecdotal and of dubious scientific merit. Can we find alternative
natural explanations?

In the absence of stupendous events such as the long-awaited Second
Coming, meticulous scientific procedure provides the only known means
by which a case for the existence of the paranormal can ever be made.
In the modern era, scientific evidence for a world beyond matter has been
sought within the confines of the laboratory. Many claims have been made,
and are still being made, of the discovery of such evidence. These have
been subjected to intense critical review. Many attempts also have been
made to independently replicate reported effects. We may ask, what is the
conclusion of these intensive efforts?

Beyond the data are the theories. The basic conflict between materialism
and spiritualism parallels a similar conflict between the points of view of
atomism and holism. In our current model of the universe, the material
world is described by discrete bits of matter that interact with one another
via the exchange of other discrete bits. The universe is organized, by spontaneous evolutionary processes, from these bits of matter. Some holistic
models, on the other hand, posit a universe featuring continuity and a
simultaneous connection between all events that, we will see, violates current
physical principles. Can we find evidence to support this concept? Is it
rational?

Other, weaker principles, usually also termed holistic, can result by
natural material means. They emerge as new phenomena, such as living

organisms and human intelligence, as matter becomes more complex. Still, these principles do not seem to derive directly from physical particles and forces.

The real issue, we will see, is not whether such principles exist—but whether they require the operation of new physical laws beyond those of microscopic physics, chemistry, and biology. Only if holistic principles have the power to violate these laws can they be termed nonphysical or supernatural.

The holistic view of many paranormal theories imagines, if not always quite so explicitly, a universe filled with a continuous field in which the bits of matter are embedded, making possible a simultaneous connection between all events in the universe. Up until this century, such a field, called the "aether," was believed to provide the medium for gravity, electromagnetism, and, some thought, other forces such as the psychic powers of the mind.

With the twentieth century revolutions of relativity and quantum mechanics, the idea of a mechanical aether disappeared from physics. But the notion of a universal cosmic fluid or aether is retained, we might say "in spirit," by modern paranormalists and New Age gurus as the medium by which the human mind interacts with a universal "cosmic consciousness."

Some of the properties of quantum mechanics, labeled as "spooky" by Einstein in 1935, have been triumphantly confirmed by experiments in the half century since. This has provided paranormal theorists with a claimed basis for both instantaneous mind-to-mind communication and the purported ability of mind to move objects (other than those, such as the eyes and hands, directly wired to the brain). Is there any basis for these claims? Can evidence for "energies of consciousness," such as human "auras," be found in observed phenomena such as Kirlian photography?

The materialist view is that, as far as we can tell from the use of the most sophisticated instruments and the theorizing of modern science, *the universe is matter and nothing more.* Still, the strong desire to believe in a world beyond exists in each of us. Possibly it is even programmed in our genes. The world of our senses is too unhappy, too unjust, too fleeting for many to accept. But we cannot let this desire for perfection and immortality cloud our judgment of the facts.

If the world is purely matter, and humans simply highly organized chunks of matter, we cannot will it to be otherwise or travel through life in an imaginary dreamworld. Instead, we must use the finest tools of our amazing intellect to determine, as best we can, the reality that is out there beyond our bodies and brains. Once armed with this knowledge, we must accept it and proceed to seek ways to live out our lives in contentment.

The world's religions and certain secular philosophies have asked humanity to submit itself to the will of invisible and possibly nonexistent transcendent forces. I propose instead that we submit ourselves to the remarkable power of the human intellect, wherever that leads. Doing so, we will recognize that each of us is ultimately free to think and act in ways that are determined by our own consciences, not the wills of gods and their priests.

Acknowledgments

This manuscript has benefited from the critical reading, in part or whole, of a number of my friends and colleagues: philosophers Ron Amundson, Paul Kurtz, Larry Laudan, and Cassandra Pinnick; physicists John Learned and Xerxes Tata; psychologists Ray Hyman and Karuna Joshi-Peters; nutritionist and health author Kurt Butler; and neurochemist Bruce Morton. I asked each to play the devil's advocate, and none hesitated to point out errors, inconsistencies, and flawed arguments. I hope these have all been corrected, and I have incorporated many of the other thoughtful comments of this varied and talented group. Of course, I take full reponsibility for any deficiencies that may remain. I am also grateful to Prometheus Books, and in particular Paul Kurtz and Doris Doyle, for their confidence in my work, and to Jack Foran for his conscientious editing. Finally, it is with gratitude and love that I acknowledge the expert critical reading for grammar, style, and argument by my life's companion, Phyllis Stenger.

1.

Supernova Shelton, Supermind, and the Supernatural

Much which we are wont to regard as solid rests on the sands of superstition rather than on the rock of nature. It is indeed a melancholy and in some respects thankless task to strike at the foundations of beliefs which, as in a strong tower, the hopes and aspirations of humanity through long ages have sought refuge from the storm and stress of life. Yet sooner or later it is inevitable that the battery of the comparative method should breach these venerable walls, mantled over with the ivy and mosses and wild flowers of a thousand tender and sacred associations. At present we are only dragging the guns into position: they have hardly begun to speak.

Sir James Frazer, *The Golden Bough*

Supernova Shelton

One hundred and sixty thousand years ago, a giant blue star in the Large Magellanic Cloud (LMC), the larger of the two satellite galaxies that reside just outside our own Milky Way, exploded in one of the most violent events the universe ever produces: a supernova. The star, called Sanduleak -69° 202 in astronomical catalogs, was eight times more massive than our own sun and fifty times bigger. After a billion years or so of life, the star extinguished all of its nuclear fuel. With nothing left to generate the pressure needed to support its immense weight, the inner core of the star rapidly collapsed, setting up shock waves that blew off its outer shell in a mighty blast. The light from the explosion, traveling 300,000 kilometers per second, did not arrive at earth until February 23, 1987.

Before the blast, the star was intrinsically a hundred thousand times brighter than our own sun. Even so, because of its great distance, Sanduleak -69°202 was invisible to the naked eye on earth. Now it became fifty million times brighter than the sun.

In the wee hours of February 24, Oscar Duhalde, a telescope operator at the Las Campanas Observatory in the Andes mountains of Chile, happened to walk outside and look in the direction of the LMC. He noticed that the Tarantula Nebula in a region called 30-Doradus was unusually bright. But he forgot about it when he returned to his duties.

On another nearby telescope, Ian Shelton, an astronomer from the University of Toronto, was photographing the LMC. After developing the film, he quickly realized what had happened. He knew exactly what to do. After alerting the other observers on the mountain, he followed standard procedures that have been set up by the international astronomical community to handle unexpected cosmic events. Shelton cabled the Central Bureau for Astronomical Telegrams in Cambridge, Massachusetts, which then relayed the message around the world: SN 1987a, the first supernova of 1987, had occurred. But this was no ordinary supernova, at least for earthlings. Although twelve additional supernova would be detected that year in other far off galaxies, Shelton's supernova was unique—the first seen in our galactic neighborhood in 383 years.

One Hundred Trillion Neutrinos

The energy radiated as visible light by SN 1987a was fifty million times that radiated normally by our sun. A hundred times more energy than this, the energy of five billion suns, was carried by the neutrinos that sprayed in all directions from the center of the blast. These neutrinos originated in a region less than thirty miles across, yet for an instant their power was greater than the radiated power of all the luminous matter in the universe (Woosley 1988)!

On that February day in 1987 when the neutrinos from SN 1987a finally reached earth, three hours ahead of the visible light that would be seen at Las Campanas, a hundred trillion of them passed through the body of each human being on earth. But the ghostly particles probably did no damage. A human body has far too few atoms. Neutrinos easily traverse the mostly empty space in between these atoms. Even the earth is almost invisible to a neutrino.

Having no measured mass and able to pass through a light-year thick wall of lead, the neutrino is often called "ghostlike." It is as close to being nothing as something can be, undetectable by the human eye and all but the largest and most elaborate particle detectors.

Neutrinos Through the Earth

When the announcement of the supernova spread around the world, three scientific groups, in Japan, the United States, and Europe, that had been searching for rare processes with massive particle detectors located deep underground, quickly ran the data tapes for that day through their computers. Although the three experiments were located in the northern hemisphere, where the LMC is never above the horizon, the experimenters knew that the neutrinos from the supernova would easily penetrate through the earth to the detectors.

All three groups immediately reported neutrino signals. The Japanese experiment, in a zinc mine in Kamioka, found eleven events. That is, of the many trillions of neutrinos that passed through their apparatus, eleven neutrinos were detected (Hirata 1987). The U.S. group, working in a salt mine near Cleveland, saw eight within the same thirteen-second period (Bionta 1987). Mount Blanc reported seven neutrinos arriving 4.7 hours earlier (Aglietta 1987).

Colleagues of mine from the University of Hawaii played an important role in the Cleveland experiment. At the time of the great event, I was on sabbatical in Rome and received a call from Hawaii asking me to help coordinate with the Italian group working under Mount Blanc. So, although I was not directly involved as an experimenter, I had an opportunity to view the excitement close-up and participate in numerous seminars and discussions. One of many issues was the discrepancy in reported arrival times of the neutrinos.

Because of this disagreement in exact arrival times and other technical questions, the Mount Blanc results were regarded as questionable. Later, a Soviet experiment under a mountain in the Caucasus would report having received a signal at about the same time as the others, but problems with determination of the absolute event times prevented the Soviets from claiming a certain confirmation. Despite these still outstanding questions, which are not unusual in science, little doubt is expressed that neutrinos from 160,000 light years away—the first cosmic neutrinos ever detected from beyond the solar system—had been observed by at least two independent groups.

Supernatural Supernovae?

If the supernova had occurred in our own galaxy, rather than the neighboring LMC, the neutrinos would have had less distance to cover and would have reached earth ages ago, passing through the earth unbeknownst to anyone. Only in the last few decades have human beings developed the technology to detect neutrinos. The visible light from the supernova, however, would have been noted by primitive skywatchers in the southern hemisphere and undoubtedly greeted with great astonishment as a bright new star.

Five supernovae within our own galaxy have been recorded in the current millennium, the last in 1604. The nearest known supernova was recorded by Chinese astronomers in 1054. Its still-expanding remnant is associated with the Crab Nebula. Within that nebula, the remainder of the original star is compressed into a ball of neutrons a few kilometers in diameter, spinning thirty times a second and emitting radio and X-ray pulses.

Many supernovae must have been seen by people in earlier ages, but an event occurring only once every few centuries would not have occurred in the lifetimes of most humans. So, when some of our primitive ancestors observed a bright new star in the heavens, they probably regarded it as an unnatural, that is, a supernatural, event.

Undoubtedly the appearance of the new star would be greeted with great awe and taken as a sign. For example, a child born at the same time might be assumed to enjoy the special favor of heaven, destined to be a king or great spiritual leader. We could be tempted to ask: was the star of Bethlehem a supernova?

Even today, many people accept the commonly occurring as natural, and assume that the rare happening must be a signal of portentous events from a realm beyond the normal. As we will see, this is an unjustified assumption. The rare and anomalous can still be normal and natural.

The Ghostly Neutrino

Neutrinos are neither rare nor anomalous—just hard to detect. Utilizing modern instrumentation deep underground or at the bottom of the sea, where most of the natural cosmic radiation has been filtered out, we have observed natural neutrinos from the sun and other stars. Under the more controllable situations at reactors and particle accelerators, experiments have now produced and detected millions of neutrinos. For over a decade neutrinos have been used to probe the structure of matter and the basic forces of the universe.

Although now understood to be one of the most important components of the universe, the neutrino was not discovered until this century. Wolfgang Pauli proposed its existence in 1930 to explain the missing energy observed in nuclear beta-decay. Enrico Fermi named it *neutrino,* which translates as "little neutral one." Fred Reines and Clyde Cowan first reported the observation of neutrinos from a nuclear reactor in 1956. The 1988 Nobel Prize for physics was given to Leon Lederman, Melvin Schwartz, and Jack

Steinberger for their discovery of the second type of neutrino in 1962.

These neutrinos have been produced "artificially" at the Brookhaven National Laboratory accelerator in Upton, New York. The first natural neutrinos were found by Reines and his collaborators working in a mine in South Africa, followed shortly by similar observations in the Kolar gold fields of India.

Natural neutrinos from the sun have been observed over a two-decade period by Ray Davis and his collaborators in the Homestake gold mine in South Dakota. Davis's pioneering work was eventually confirmed by the same Japanese group that detected neutrinos from Supernova Shelton. And finally, many scientists are now working on projects to develop the first neutrino telescopes able to observe very high energy neutrinos from the cosmos. Currently, I am the Deputy Director of one such project, called DUMAND: the Deep Underwater Muon and Neutrino Detector. We plan to place instruments for neutrino detection at a depth of 4.8 kilometers in the ocean off the coast of the island of Hawaii.

Stars are not the only sources of cosmic neutrinos. Very low energy neutrinos left over from the Big Bang, the violent explosion that began the universe, are now believed to uniformly pervade all of space, outnumbering the atoms of ordinary matter in the galaxies, stars, and planets of the universe by a billion to one.

Although we haven't figured out how to detect these relic neutrinos, and did not even suspect their existence until recently, we are almost certain they are there. The background of microwave photons discovered in 1964 had been predicted as a remnant of the Big Bang; about the same number of neutrinos should exist as part of that remnant.

Supersense

You may be wondering what this long song and dance about supernovae and neutrinos has to do with the theme of this book, which is the issue of the existence of a nonmaterial world. Let me try to explain.

Our normal senses respond to the world of matter. Evidence for a world beyond matter is usually presumed to be derived from other channels of knowledge available to the human mind, channels that do not rely on the data from sight, sound, smell, taste, and touch.

Mystical traditions of all ages have claimed the existence of some additional channel—the sixth sense. Furthermore, in the mystical view, the sixth sense is superior to the other five. It is a "supersense" that provides information about a deeper, truer, transcendent reality. A major part of the task I have taken upon myself is the critical examination of the evidence for supersensory sources of knowledge about the universe.

As for neutrinos, we have seen that they are a billion times more common than the chemical elements that compose the matter of normal experience. Now, you would think that, if humans actually possess a superior sixth sense, such an important component of the universe as neutrinos would have been noted ages ago.

If the human mind has a channel to fundamental knowledge about the universe that is superior to the five senses, how could that channel have missed the 10^{88} neutrinos pervading all of space or the trillions of neutrinos passing through the human body during a supernova? These copious objects, all around us but not detectable by any of the five normal senses, lay undiscovered until we developed the necessary technology.

In fact, when you think of it, why is it that the sacred literature and oral traditions, reporting the mystic revelations of those who claimed to possess this special channel to the truth, have never predicted the existence of any of the vast range of phenomena found by modern scientific instrumentation?

Search as you may through the traditional stories of all cultures, sacred or secular, you will find no mention of quarks, DNA, plate tectonics, or quasars. If visionaries have lived who have seen things that ordinary people could not, why did none of them get even the most basic facts about nature correct: about the evolution of life or the place of the earth in the universe?

Rather, these so-called visionaries got just about everything wrong. Uniformly, they placed humankind and the earth at the center of a universe populated with angels and demons, with the human being occupying a special place just below angels. This picture bears no resemblance to that which science has provided, with the earth revolving around a relatively minor star off in the corner of a relatively minor galaxy of a hundred billion stars, in a universe of a hundred billion galaxies, and *homo sapiens* no more than one particulary successful mammalian species.

The revelations in sacred books look more like the product of vivid imagination, superstition, and wishful thinking than any factual representation of the real world. If sacred revelations were so wrong about this world, how can people still believe that they tell us anything about other worlds, worlds beyond the senses, and ultimate reality?

Natural or Supernatural?

To primitive peoples, each rock and tree was a spirit, and no real distinction was made between the natural and supernatural. The division of events into natural and supernatural came with the development of science. As science progressed, phenomena were reclassified one-by-one from the realm of the supernatural to that of the natural. The abnormal became normal, once natural explanations were found. Feats that were once regarded as impossible to humans, such as aerial flight or communication across great distances, became commonplace.

Historically, the term "supernatural" has been applied to phenomena that are unusual, anomalous, and beyond conventional explanation. However, in science today, anomalousness or abnormality are hardly evidence of the supernatural. An adequate natural explanation for every reported mysterious or unusual event has always been found, if sufficient data on the event existed. In fact, science thrives on the abnormal, searching out anomalous events as roadsigns marking the way to new discoveries about nature, not supernature.

The Anomalous and the Unexplained

Several of the laboratories that engage in psychic research use the word "anomaly" in their names, and many of the arguments made in support of the existence of psi and other paranormal phenomena rest on observations labeled "anomalous" by the discoverers and their supporters. The presumption is that anomalies have been demonstrated to exist in psychic phenomena. In this book, I challenge that presumption.

In my view, paranormal researchers use the term anomaly in an unconventional and highly misleading fashion, applying it to phenomena that should better be termed "unexplained." The difference? Anomalousness, in conventional science, implies something new, something beyond existing knowledge. The unexplained, on the other hand, can be perfectly normal but simply lacking in sufficient data to pin down the explanation.

As an example, consider Unidentified Flying Objects (UFOs). Most reported sightings of strange aerial phenomena have simple, normal explanations. A few remain unexplained, "unidentified." UFOlogists argue that this residual set of unexplained phenomena constitutes an anomaly sufficient to indicate that something strange is going on. Skeptics, on the other hand, point out that since the data are insufficient to rule out all normal explanations, we cannot conclude that an anomaly exists.

Examples of valid anomalies, challenging conventional knowledge, abound in the history of physics. Each helped to point the way to dramatic changes in our views of the universe. Galileo's telescopic observations of the moons of Jupiter were anomalous to the geocentric model of the universe. So were his observations of the phases of the planet Venus. These and other anomalies led Galileo to adopt the Copernican model. The nineteenth-century observations of spectral lines in the light emitted by high temperature gases were anomalous to the wave theory of light. Attempting to explain these anomalies led Niels Bohr to the quantum theory of atoms. Anomalies have always played a crucial role in the development of science, and scientists constantly seek them out.

What Is Anomalous?

Robert G. Jahn, director of the Princeton Engineering Anomalies Research (PEAR) program established in 1979 at Princeton University, should know something about anomalies. In his book *Margins of Reality,* written with Brenda J. Dunne, Jahn uses a similar metaphor to the one I independently used above. Jahn and his coauthor describe how anomalies "stand like road signs, signalling 'detour' or 'wrong way' to the scientific traveller" (Jahn 1987, p. 4). We agree thus far, but not much farther.

Jahn and Dunne proceed to call catastrophic events such as "wars, plagues, economic collapses, political or religious upheavals" anomalous. Is war anomalous? Is political upheaval anomalous? Is breathing anomalous?

Later in this book we will evaluate the results from Jahn's laboratory that he claims are anomalous and thus evidence for psi. Here I want to focus on Jahn's view of what constitutes an anomaly, because it seems to reflect the philosophy expressed in much parapsychological literature— a view I claim is contrary to that of the conventional scientific community.

Besides catastrophic (but perfectly normal) events, Jahn and Dunne mislabel as anomalous many anecdotal human experiences reported through the ages as inexplicable as evidence of magic, miracles, sorcery, gremlins, and psychic phenomena. The label would be fair if the authors explained that they simply use the word anomalous as a synonym for "mysterious" or "unexplained," but this is clearly not what they mean. To Jahn and Dunne, an anomaly "will point to a more penetrating path than that previously followed, along which the science can resume its progress more vigorously than before" (p. 5).

Is science to change direction each time some catastrophic or unexplained event takes place, every time someone reports a gremlin? Science doesn't change direction unless it has a good reason for doing so. The anomalies that redirect the path of science must be profoundly in disagreement with existing knowledge, not simple everyday happenings that folklore calls mysterious.

Jahn and his colleagues in parapsychology have so far not succeeded in winning acceptance by the rest of the scientific community of their claims of anomalous psychic events. As we will see, they have failed so far to show us any real anomalies: ones that cannot be more economically explained; ones that require the introduction of revolutionary new hypotheses about the fundamental nature of the universe.

No Current Anomalies

Throughout this century, the observation of anomalies in experiments with atoms, particles, and light pointed the way to the development of the theories that now form the accepted structure of physics. However, today we are in a remarkable, unprecedented, and undoubtedly temporary position: *there are currently no fundamental empirical anomalies* in our knowledge of the basic nature of matter. Every bit of data gathered to this date in scientific laboratories around the world is consistent with the Standard Model, the existing theory of the basic nature of matter. The details of the Standard Model will be presented in the next chapter. But those details are not yet needed for my discussion. Here I am simply trying to clarify the concept of anomalousness.

What do I mean when I say there are no anomalies in our knowledge of the basic nature of matter? First let me tell you what I am *not* saying. I do not claim that we know everything. Certainly we cannot calculate in detail all the processes involved at every level of observation, and we still have an enormous amount to learn about the universe. For example, at this time we do not have a satisfactory explanation for the marvelous phenomenon of high-temperature superconductivity or for the formation of galaxies. Great unanswered questions remain in all sciences, and I am sure revolutionary discoveries will be made in the future, just as they have in the past.

Despite the success of the Standard Model, many nagging questions persist; but these are of a theoretical rather than empirical nature: the problems are with the theory, not with the experimental facts. The Standard Model is unsatisfying to the people working on it, yet the model itself still works; it successfully describes all the currently observed properties of the known fundamental particles and processes, in other words, all existing data.

Occam's Razor

My argument here is grounded on the principle referred to as "Occam's razor," or the law of parsimony. Occam's razor is a tool of rational thinking that forbids us from introducing any hypotheses other than those required by the data. In choosing between theories of equal explanatory power, we must discard those that are least economical.

We can speculate all we want, and indeed we need such speculation to stimulate new thinking. But an hypothesis cannot be retained simply because it is not ruled out by experiment, or because it could be true, or even because it is the best fit to the data among all competing hypotheses. Supporting evidence must be found to show why we cannot do without the new hypothesis. And doubters are not required to prove that the proposed hypothesis is wrong. The burden of proof is on the proponents of a new hypothesis.

For example, the existence of forces weaker than those so far observed by physicists is not ruled out. But proposals for such forces, or new fundamental particles, will remain speculative until some empirical data are found that require adding greater complexity to the existing picture.

One can always invent complex, after the fact, ad hoc theories of nature that "explain everything." Many people do this, and can't understand why the science establishment ignores them and rejects their submissions to the science journals. The reason for these rejections is Occam's razor.

Accepted scientific theories are usually the simplest we can come up with, those having the fewest assumptions. In our current understanding of the particles and forces of nature, no fundamental hypotheses are needed beyond the Standard Model. So don't bother to send your alternative model in to the *Physical Review* unless it explains a new piece of data not already covered by the Standard Model, or you can show that your model is simpler than the Standard Model, that is, that it is based on fewer assumptions. It's up to you, not the editors or reviewers, to show where your new theory can be of some use.

The use of Occam's razor, along with the related critical, skeptical view toward any speculations about the unknown, is perhaps the most misunderstood aspect of the scientific method. People confuse doubt with denial. Science doesn't deny anything, but it doubts everything not required by the data. Note, however, that doubt does not necessarily mean rejection, just an attitude of disbelief that can be changed when the facts require it.

If what I write here survives into the next century, and new mysteries are uncovered in the interim (as they undoubtedly will be), I can just imagine critics pointing to me as another of those closed-minded scientists who always think they have all the answers. Let me try to make it as clear as I possibly can: We do not have all the answers. Rather, we currently have no questions—at least no questions about the fundamental nature of matter forced upon us by observations of the real world that threaten the Standard Model.

Certainly we have many practical questions, theoretical questions, and speculations, but nothing in the data gathered by our finest instruments yet points toward hypotheses that would diverge from the Standard Model. This situation is not likely to remain. Surely new anomalies will be found. Anomalies are fun; they are welcomed with open arms by research scientists because they give us interesting problems to solve.

The new anomalies, when they are found, will undoubtedly result in the rejection of the current Standard Model. Possibly they will even lead us to revoke the materialist, reductionist, and quantum mechanistic view of the world that now works so well. But if this happens, it will be because empirical evidence demands it, not simply because of pious philosophizing or wishful fantasies based on the superstitious beliefs of the prescientific age.

Supernaturalism

For reasons of scientific ignorance, cultural tradition, and psychological needs, most people today still believe in some aspects of the supernatural. Traditional religions are founded on the concept of supernatural forces responsible for virtually everything, from the origin of the universe to human behavior to making a leaf fall to the ground.

Supernatural belief continues to be strong in almost every part of the world, curiously gaining strength most recently in those varieties referred to as "fundamentalist," which are in the greatest conflict with science. The fundamentalists dispute the first tenet of science: that authoritative knowledge must be based on objective observation and rational analysis. They insist on the authority of sacred scriptures.

In addition to this reheating of the ancient conflict between science and religion—which I believe are fundamentally incompatible—we are now also confronted with a modern "secular supernaturalism." The occult has been a part of human life since the ancient days of the shaman. Astrology is still practiced much as it was by the Babylonians. Alchemy is largely abandoned, but one still hears occasional stories about modern witchcraft. The spiritualism of the last century has evolved into the psychic and psychedelic phenomena of this century. Today, the newest form of supernaturalism, collecting under its umbrella many traditional and modern occult ideas, is called the "New Age."

Supermind

The common thread of both ancient and modern supernaturalism, religious and secular, is belief in the Cartesian split: the duality of mind and matter. Unable to deny the evidence that our bodies are weak and mortal, many of us cling to the desperate hope that we have immortal "superminds," that we are part of some great "cosmic consciousness." Again, the basic idea is that our minds possess channels involving dimensions beyond time and space.

One might argue that support for the concept of supermind can be found in the very success of science itself, especially in the remarkable way that mathematics can be used to represent real phenomena. Any student of physics cannot help but be astonished by the power of a few mathematical equations, the product of human thought, to predict events convincingly and with great precision.

Newton's laws, Maxwell's equations, Einstein's special and general theories of relativity are all just scribbles on a chalkboard, yet out of them come predictions about future events far more reliable than those of any ancient seer, prophet, or contemporary fortune teller operating under the guise of psychic or channeler. How does it happen that scientists can achieve these wonders? Isn't this supermind?

In many cases, mathematical theories are developed with little or no apparent direct empirical input. Sometimes they are motivated by an argument of mathematical elegance or symmetry. For example, in the latter part of the nineteenth century, James Clerk Maxwell (1831–1879) made a small modification to the set of equations that had been developed by several predecessors to describe the properties of electricity and magnetism. Maxwell's modification was motivated, at least in part, by an esthetic feeling that the equations for the two phenomena should possess a certain symmetry. When Maxwell then applied his new symmetrical equations to an experiment within a vacuum, where no electrical charges or currents existed, he discovered that the vacuum still could contain electromagnetic waves.

Maxwell computed that electromagnetic waves in the vacuum would travel precisely at the speed of light ($c = 300,000$ kilometers per second). This quantity was not directly entered into the calculation, but magically seemed to come out of it. Here was an empirical quantity, determined by measurements in the laboratory, that seemed to have been derived from purely mathematical and esthetic arguments.

Maxwell's result suggested that light was an electromagnetic wave and predicted that it should be possible to generate and detect radio waves. This prediction was confirmed in 1888 by Heinrich Hertz. So our modern world of telecommunications is to a great degree the product of mathematical esthetics.

Kantian Pure Thought

Undoubtedly the equations of physics testify to some remarkable power of the human mind. Are they not examples of Kantian "pure thought," providing a channel to truth independent of observation? Isn't the very success of mathematical science evidence for the mind being tuned into some "supersensory" source of information? Can't this be interpreted as support for the basic claim of mythology, that the deepest truths about ourselves and the universe lie within?

I claim that, deep as the truths obtained by reason and insight may be, their original source of data is still sensory. Kant argued that the axioms of geometry were the result of pure thought, yet they were undoubtedly inferred from observation. The early Greeks probably figured out their rules for parallel lines and triangles from lines and figures drawn in the dirt with a stick.

It eventually turned out that other axioms could be used to define different geometries that may or may not have anything to do with the real world. These non-Euclidean geometries were only mathematical curiosities until Einstein discovered that they provided a profound alternate way to describe gravitation.

Still, Euclidean geometry remains a good approximation for most of space and appears to be the correct average for all of space. This could explain why our knowledge of certain laws of geometry appears to be innate. But even if this is so, if this knowledge is innate, it could have been built into our genes by the experiences of our ancestors. Spatial thinking may have had practical survival value.

Similarly, the speed of light calculated by Maxwell in fact originated in sensory observations. The value $c = 300,000$ kilometers per second obtained from Maxwell's equations results from two other empirically determined quantities, constants measured from experiments in electricity and magnetism. And this prediction is checked by another observation, the directly measured speed of light.

Maxwell's equations, general relativity, and the other great theories of physical science are derived from the analysis of sensory data, looking for patterns that might go beyond current observations. They are basically extrapolations of observations.

We perform similar extrapolations in everyday life. For example, after experiencing a rainy day, we take our umbrellas to work the following morning on the reasonable chance that it will also rain that day. And if it does rain, we do not assume we possess some great psychic ability to see the future.

Machines are not Human

Carl Gustav Jung (1871–1961), the founder of analytical psychology, proposed that myths are part of the collective unconscious of humankind, the supposed repository of universal spiritual truths. We seem to be born with certain moral conceptions, although this is a subject of continual debate. However, even if we do possess such moral knowledge at birth, it could have been programmed in our genes by the processes of evolution, arising originally from the experiences of our ancestors. We have no basis for believing that this knowledge comes from a realm beyond sensory reality.

Now, we can be fooled by our direct sensory observations, and we can delude ourselves with subjective interpretations of sensory data. Sensory data must be rationally and objectively analyzed before they can be adequately interpreted. For example, our immediate experiences tell us that the world is flat. But our reason, analyzing and interpreting data from these and other observations, tells us: (1) the apparent flatness could result if the earth is sufficiently large; and (2) roundness can explain many other observations, such as ships disappearing over the horizon in one direction and reappearing months later on the horizon opposite. We trust our senses if the conclusions based on their data make sense.

Science is the activity of collecting observational data and applying rational analysis to these data. Mathematics and logic provide us with analytical tools, but equally important are the instruments that enable us to greatly extend our senses. These instruments have peered deeply inside the nuclei of atoms, and far out into the universe, and found no sign of supernatural forces.

But, it still is argued, these machines are not human. Not being human, perhaps our inanimate scientific instruments are insensitive to uniquely supernatural phenomena accessible only to the minds of human beings.

Human nature leads us to believe that we are not just another species of animal or assemblage of quarks and electrons, but a transcendent super-creature whose true milieu exists somewhere between the radiant glory of heaven and the slimy mud of earth. If the mind really can penetrate beyond matter via a supersensory channel to a cosmic consciousness, then we would not expect machines, which are merely material, to find anything.

This should come as a relief to all hard-pressed taxpayers, since the machines in question, such as giant particle accelerators and space-based telescopes, cost many millions of dollars. The search for the existence of a world beyond the senses would certainly be of sufficient importance to justify great public expenditure to gain the priceless evidence. But, thank goodness these millions are not necessary, since the apparatus we need to gather the evidence purportedly already lies within the immaterial soul of each of us.

Does Thinking Make It So?

The idea that a world exists beyond the senses, accessible through some sort of mental sixth sense, and that human beings can harness the power of that world to move mountains, overthrows many physical laws that have been built up on the basis of observation over centuries. Basically, psi is miraculous. If it exists, it violates well-established principles of physics: the laws of motion, conservation of energy, causality, and relativity.

In the scientific view, every body in the universe is made up of particles moved about by physical forces, with an overlaying superposition of randomness making it impossible to predict the outcome of everything that happens and implying that much of what does happen is undetermined—the result of chance. The supersensory proposition, on the other hand, says that these bodies can be moved by mysterious powers: spirits or the mind itself.

This is not a new idea. The belief in underlying invisible forces interacting with the observed world has been widely held since people first thought about such things. It began with primitive animism, in which everything was thought to be alive and moved by invisible personified spirits. Over the millennia, these ideas developed into more abstract concepts of God and soul that are still widely held today. People ask: Would these beliefs not have disappeared long ago if there were no truth behind them? Aren't they common sense?

Someone has defined common sense as the human faculty that provides us with basic knowledge about the world, such as that the earth is flat. Belief in a flat earth is just as ancient and universal as the belief in spirits, but this did not make the earth flat. The sun does not go around the earth, just because millions of human beings throughout the ages thought it did. And neutrinos exist, even though no one until this century thought of them.

An Alternative

If phenomena beyond the evidence of the five senses, or scientific apparati, have not yet been found, nor evidence for super powers of the mind transcending known physical principles, then why is belief in the supernatural so prevalent? Even the new secular spiritualism offers that traditional promise of religion: Give me money and I'll help you live forever. Philosopher Paul Kurtz has named this the "Transcendental Temptation" (Kurtz 1986). Selling eternal life is an unbeatable business, with no customers ever asking for their money back after the goods aren't delivered.

Some people don't like to hear what science tells them: that humanity is an insignificant speck in a universe that may have spontaneously appeared without cause or plan, and that when you're dead you're dead. But many others prefer this to the product that the purveyors of supernaturalism have to offer: the notion that we are all puppets dancing on the ends of supernatural strings; that all the great achievements of the human race— art, literature, science, morality, democracy, and justice—are not really a result of our own efforts but handed down from above.

Those who have no need to deny the reality they see with their own eyes willingly trade an eternity of slavery to supernatural forces for a lifetime of freedom to think, to create, to be themselves. The wonders of this world are more than sufficient, and the powers of our minds more than adequate for us to enjoy these wonders.

References

Aglietta, M., et al. 1987. *IAI Circ.* 4323; *Europhys. Letters* 3:1315 and 1321.

Bionta, R. M., et al. 1987. *Physical Review Letters* 58:1494.

Hirata, K., et al. 1987. *Physical Review Letters* 58:1490.

Jahn, Robert G., and Dunne, Brenda J. 1987. *Margins of Reality: The Role of Consciousness in the Physical World.* New York: Harcourt Brace Jovanovich.

Kurtz, Paul. 1986. *The Transcendental Temptation: A Critique of Religion and the Paranormal.* Buffalo, N.Y.: Prometheus Books.

Woosley, S. E., and Phillips, M. M. 1988. "Supernova 1987A!" *Science* 240:750.

2.

The Standard Model and the Search for Anomalies

However we select from nature a complex [of phenomena] using the criterion of simplicity, in no case will its theoretical treatment turn out to be forever appropriate (sufficient). . . . I do not doubt that the day will come when [general relativity], too, will have to yield to another one, for reasons which at present we do not yet surmise. I believe that this process of deepening theory has no limits.

Albert Einstein

The Simplest Picture

If I were to summarize as succinctly as possible the view of this book, it would be as follows:

At this writing, neither the data gathered by our external senses, the instruments we have built to enhance those senses, nor our innermost thoughts require that we introduce a nonmaterial component to the universe. No human experience, measurement, or observation forces us to adopt fundamental hypotheses or explanatory principles beyond those of the Standard Model of physics and the chance processes of evolution.

In this chapter, I present in some detail the current picture that physicists have of the material universe at its most basic level of structure. The "fundamental hypotheses or explanatory principles" referred to in the above summary are collected in a composite of physical theories called the Standard Model. This model successfully describes all that is empirically known about elementary particles and the forces through which these particles interact.

I claim that nothing about the universe observed to this date—microscopic and macroscopic, animate and inanimate, human and nonhuman— shows any anomalous behavior that is inconsistent with the simplest picture of a universe composed of matter assembled from these particles. The hypotheses of other, nonmaterial components to the universe are not justified by the data. While we can never prove that such components of the universe do not exist, we cannot logically include them. They are currently forbidden by the law of parsimony.

Have I reduced everything to particle physics? Not quite. Using the highest energy accelerators and most sensitive detection equipment that current technology is capable of producing, particle physicists attempt to uncover the elementary objects that compose the universe. Nuclear, atomic, and molecular physicists investigate the structures produced when these quarks and leptons combine. Other physicists, such as those in the wideranging condensed matter specialties, study the remarkable physical properties that result when atoms and molecules assemble in various complex configurations in solids, liquids, gases, and plasmas.

Chemists investigate the countless substances produced when atoms combine into molecules. The biochemist gets into the act when these molecules are part of the stuff of life. And so it goes, on up the ladder, as various scientific disciplines examine the systematics of the complex structures that occur as matter is arranged in a variety of configurations.

Few scientists need to know much about particle physics to pursue their goals. They observe their own subject matter directly in the laboratory or field, and determine its useful orderly properties, which they then describe in their own specialized jargon. They don't derive these properties from the Standard Model of particle physics; nor do they describe their results in the language of quarks, leptons, and gauge bosons.

Emergent Structures

Where did these orderly properties, observed at every level of human experience, come from? The determined reductionist would say that they still arose out of quarks, leptons, and gauge bosons; we simply are not yet smart enough to calculate how. Near the other end of the spectrum are the holists, who say that the higher levels of organization observed for matter of increasing complexity are evidence for some still undiscovered set of cosmic principles that operates across space and time, unfettered by any bonds implied by the microscopic laws of particle physics.

The extreme end of this side of the spectrum, at the furthest remove from the uncompromising reductionist, consists of those who see the material universe as an illusion of transcendent cosmic mind. However, someplace near the center we find people who take the view that the properties observed for the complex material systems are a result of a perfectly natural mechanism of self-organization that occurs whenever material bodies aggregate into more complex systems. These so-called "emergent properties" are fully material, yet not totally reducible to quarks and leptons.

The questions that will themselves emerge as this book progresses can be summarized as follows: (1) Does evidence exist within the macroscopic objects of our experience for any special holistic principles not derivable from elementary particles and forces? (2) If these holistic principles exist, can they be interpreted as the natural operation of known material forces? (3) Can the properties we describe as life and mind be understood without invoking nonmaterial principles? Even if the answer to the first question is yes, a yes answer to the other two will leave no room for a spiritual, supernatural, or paranormal component in human experience.

By Eye and by Instrument

But before we can tackle these difficult issues, we need to understand the basic picture particle physicists have developed that describes, without any current empirical anomalies, all that has been observed in the subatomic realm: the Standard Model. We will see how the Standard Model forms a basis for our understanding of the larger everyday world of common experience, as well as of the incredibly more vast astronomical universe. Indeed, we will be able to push back to the beginning of time and speak about how the universe itself could have happened spontaneously, without the intervention of a creator or other supernatural act.

I will not attempt to describe all the evidence accumulated over the last few decades in support of the Standard Model itself. That would take many pages, indeed many books, and this evidence is not at issue.

Nonetheless, we must keep in mind that the Standard Model is based first of all on empirical data. While we cannot rule out the possibility that other equally successful models might be invented in the future, or even suggest that the model will continue to hold up when new empirical realms are discovered, the Standard Model is the simplest picture we have been able to draw that is consistent with the multitude of data so far accumulated by eye and by instrument.

The amount of data is vast, but the model that so successfully describes that data is amazingly compact. Nature seems to be complex in detailed structure, but very simple in essence. As we will see, the beauty and complexity of life on earth are not the result of complex principles. Rather they result from an even more beautiful simplicity.

At the outset, the Standard Model is firmly grounded in the atomism first ennuciated, in ancient times, by Thales, Empedocles, Democritus, Epicurus, and Lucretius—but then almost forgotten for many centuries until resurrected 400 years ago by Newton and others, and finally brought to full flower in this century with the empirical establishment of the particulate nature of matter and light.

Matter

In both ancient and modern versions of atomism, the entire universe, indeed every object of our experience, is composed of discrete matter. To the early Greek atomists, even the soul was matter. Physicists define matter as substance that has "inertia," that is, that resists changes in its motion. The greater the inertia, the harder it is to get it moving, or to stop it when it is moving. The physical quantity associated with the inertia of bodies is called "inertial mass." This is the mass we talk about when we say a body has a mass of one kilogram.

At one time it was believed that the universe was made of two kinds of stuff: concrete matter, as exemplified by rocks, trees, and water, plus something more ephemeral and wispy, like air or light. But even ancient thinkers, notably Lucretius, realized that air was inertial in nature, as evidenced by the power of wind to move objects. Air has inertia, and so it is matter.

Light, however, seemed different to early observers. For one thing, light did not seem to possess the inertial properties we associate with massive bodies. In normal experience, light beams do not move objects. We can feel the wind pressing against our bodies after coming out from behind a barrier, but no similar experience occurs in moving from shade to sunlight.

Light is Matter

Although we cannot feel the pressure of sunlight, we can feel its heat. This suggests that light carries energy, because we associate heat with energy. This, in turn, leads us to conclude that light is a substance different from matter, a form of pure energy. And so the distinction between energy and matter arises: two unique forms of the stuff of the universe.

This distinction turns out to be artificial and unnecessary. Light and other forms of so-called pure energy have all the inertial properties we associate with matter. The inertia of light is normally too small to be noticed.

For example, the pressure of sunlight is a hundred billion times lower than atmospheric pressure.

Nonetheless, light falling on a sufficiently small body will cause it to recoil backward a measurable distance. This effect is called "radiation pressure." In space, the radiation pressure of the sun is a kind of a wind that blows particles out of the solar system. The radiation pressure of sunlight may even be harnessed someday to enable spaceships to sail between the planets.

The inertia of light can be understood from Einstein's theories of relativity. In the special theory of relativity, first formulated in 1905, the energy (E) and the inertial mass (m) are shown to be simply related by the well-known equation $E = mc^2$, where c is the speed of light. In the general theory of relativity, developed ten years later, Einstein asserted the equivalence of inertial and gravitational mass. Furthermore, he determined the precise interaction between light and gravity and predicted the bending of light by the sun. The quantitative confirmation of this prediction was one of the triumphs of Einstein's theory.

Energy is Matter

Einstein showed that energy and matter are the same stuff. One of the primary hypotheses of relativity is that the speed of light (c) is a universal constant independent of the motion of the source or observer. So, Einstein's equation $E = mc^2$ establishes the equality of E and m. The factor c^2 just makes a trivial change in units, for example, from kilograms of mass to Joules of energy in the Standard International metric system of units. Except for this baggage, an archaic concept carried over from previous centuries, we could simply write $E = m$. Energy and inertial mass are, for all purposes, the same.

In 1905, Einstein extended the original quantum idea of Max Planck and proposed that light is composed of particles he named "photons." These photons carry energy from one point to the other; and since energy is mass, photons have mass. Some confusion may exist here because of the common statement that photons have zero mass. This refers to rest mass: the mass/energy particles have when they are at rest. Normally, when physicists use the term mass, they mean rest mass. However, the mass I speak of here is the inertial mass of both Newton and Einstein: the property of matter that resists changes in motion.

Inertial mass is the defining property of matter, so light and every other form of energy are matter. Astronomers and cosmologists often distinguish between two components of the universe: matter and radiation. But this distinction is also artificial and unnecessarily confusing. The astronomer's radiation is just matter moving at speeds near the speed of light. All known radiation is matter. And so, the various forms of nuclear radiation—alpha, beta, and gamma—are also material.

Spiritual Energy

We still hear the term "energy" used to describe certain phenomena in such a way as to distinguish these phenomena from the grubby everyday business of ordinary matter. We read about "spiritual energy" or "mental energy," as if the use of a physics term for a concept thereby guarantees it scientific validity. But since energy is matter, it is limited by definition to the material world.

One of the ironies and irrationalities of common antimaterialist rhetoric is the way terms like "energy" and "resonance" are used to describe non-material concepts. These are physical concepts, operationally defined in terms of matter, with no logical meaning beyond that realm. Even space and time are undefined outside the material realm. If a nonmaterial world exists, then we should describe it with words appropriate for that world, rather than words that are precisely defined only for the world of matter. The concepts associated with words like soul, spirit, and supernatural supposedly transcend the space, time, matter, and energy of the material world.

Photons, Neutrinos, and Dark Matter

Matter comes in many forms. Since chemical elements such as oxygen, nitrogen, carbon, and iron compose a major portion of the material of common experience, we might be led to think that these elements are the most plentiful in nature. But this subjective perception is wrong. Astronomy books tell you that the universe is 90 percent hydrogen. But this is wrong too. In fact, photons and neutrinos are a billion times more plentiful than the hydrogen and other elements in all the stars and galaxies.

Approximately 10^{87} photons and neutrinos are estimated to uniformly fill all of space. They are the expanding remainder of the Big Bang, the explosion that began the universe fifteen to twenty billion years ago. The remnant photons started out as gamma rays one million times as energetic as visible light. Now they have cooled by expansion to just three degrees above absolute zero. They now occupy the microwave region of the electromagnetic spectrum and have less than a thousandth of the energy of the photons of visible light. These microwave photons, which apparently uniformly pervade all the universe, were first detected by Penzias and Wilson in 1964. The remnant neutrinos remain undetected, but we have no reason to doubt their existence.

Not only do the chemical elements compose only a tiny fraction of the universe by particle number count, they also compose only a small fraction of its mass. We have evidence for another yet-unidentified component, possibly as copious in number as photons and neutrinos, that provides 90 to 99 percent of the mass of the universe. This is the so-called dark matter.

The evidence for this unseen component is indirect, but strong, nonetheless, because it is based on well-known principles of gravity. For example, the visible stars within galaxies do not move in the orbits expected based on the visible matter in that galaxy. Rather, the stars appear to be orbiting within a smooth distribution of matter that extends well beyond the visible limits of the galaxy.

The discovery of unseen objects by means of their gravitational effects on other visible objects has a long and glorious history in astronomy. The planet Pluto was discovered through its perturbation of the orbits of Uranus and Neptune. Now, of course, Pluto can be seen with sufficiently powerful telescopes, confirming its existence and providing support for my contention that science offers our best bet to find evidence for the invisible.

Dark matter appears to be the dominant gravitational component in and between the galaxies. The stars and galaxies, which occupy most of the attention of astronomers, play a small role in determining the basic dynamical structure of the universe. However, the exact nature of the dark matter is a great unanswered puzzle at the current time, and major efforts are underway to solve it.

A Dark Anomaly?

We know much about the universe, but still cannot identify its most important component. Then, you might ask, why don't I label dark matter an anomaly? My answer is that we have no data forcing us to conclude that the solution to the dark matter puzzle lies outside of conventional theory. Indications exist that the dark matter is not ordinary stuff like dead stars, planets, rocks, or protons, although it may be massive neutrinos. Perhaps the dark matter, when it is identified, will turn out to be anomalous. I hope so! As I have emphasized, we are somewhat frustrated now by the lack of any anomalies in nature.

So the universe is mostly microwave photons, neutrinos, and dark matter. None of these components is visible to the naked eye, demonstrating the paucity of information that unaided human sensory observations provide about the universe. Modern science began when Galileo turned his telescope on the heavens and saw that the earth was not the center of the universe. Before we learned to build instruments, we knew almost nothing. All the folklore and mystical revelations of centuries gave us no hint of the universe now revealed by our instruments.

The Unburned Remnant of the Big Bang

The chemical elements that make up the stars and planets are a tiny unburned remnant of the Big Bang, like the residue left at the bottom of the test tube after the rest of its contents has gone up in smoke. The 10^{79} particles left over from that gigantic initial explosion that now forms the visible universe are largely collected in atoms. About 90 percent of these are hydrogen, as the astronomy books say, and the remainder mostly helium. Less than one percent of the remainder comprises the elements beyond helium in the Periodic Table. Most humans concern themselves exclusively with this tiny portion of reality. But we grandly delude ourselves when we think that the everyday experiences of puny earthlings have much to do with the fundamental nature of the universe.

Quarks and Electrons

The atoms that make up stars, planets, rocks, trees, and humans are themselves composite objects. Each consists of a cloud of electrons surrounding a nucleus several thousand times smaller that, nevertheless, contains almost all the atom's mass. The nucleus is composed of protons and neutrons, particles now known to be composite objects themselves, formed from two types of more elementary objects called "quarks."

In a somewhat oversimplified view of the Standard Model, but adequate for our purposes, the proton (p) consists of two "up" quarks (u) and a "down" quark (d), or p = uud. The neutron (n) consists of one u quark and two d quarks, or n = udd. The primary particles of the universe are thus the following: photons, u and d quarks, electrons, and neutrinos.

Incredible as it may seem in its simplicity, our current picture is that these five fundamental particles compose virtually all the known universe. While we now reserve the term "atom" for the composite quark-electron structures that form the chemical elements, our picture of the basic nature of matter remains the ancient atomic one: discrete particles constitute the universe. The popular holistic notion of a continuous universe, with everything acting together in simultaneous harmony, bears no resemblance to the Standard Model.

Other Particles

Beyond normal human experience, however, nature is not limited to the five fundamental particles mentioned. Our instruments reveal that, for brief periods of time, other more ephemeral particles appear that disintegrate within tiny fractions of a second, leaving as remnants the five stable particles.

For example, a heavy version of the electron, the "muon," is produced when high energy protons and other nuclei in the cosmic rays hit the earth's atmosphere. Muons live for only a few millionths of a second, but move fast enough to reach the surface of the earth in their brief lifetime, constituting the major component of cosmic radiation reaching the surface. Approximately one muon produced by particles from outer space passes through each square centimeter of your body each minute—another important fact only scientific instruments reveal.

A host of other even shorter-lived particles are observed, both in cosmic rays and earthbound particle accelerators. They have names like pions, kaons, hyperons, omega, rho, psi, and chi. Although these particles do not form a major component of the universe, their properties, determined from extensive investigation over the past three decades, have provided the primary empirical basis for the development of the Standard Model.

Only after large numbers of new particles were produced and studied, particularly in particle accelerator experiments, did the pattern of the Standard Model emerge. We probably would not have inferred the structure of the model from a knowledge of stable particles alone. Such an inference would be like trying to guess, from a single five-card hand, the full sequence of a fifty-two card deck. Or in a more scientific analogy, it would be like trying to determine the chemical Periodic Table from a study of just five elements.

In addition to the muon, a third electron-like particle has been confirmed that is even heavier: the "tauon." These three fundamental particles—electron, muon, and tauon—are grouped in a class called "leptons." The neutrino is also a lepton. In fact, for each of the three electron-type leptons, we find that a separate and distinct neutrino exists. We call these the electron-neutrino, the muon-neutrino, and the tau-neutrino.

The u,d doublet of quarks has been similarly joined by two additional heavier quark doublets: c and s, for "charmed" and "strange," and t and b, for "top" and "bottom" (or sometimes, "truth" and "beauty"). We should note that at this writing direct evidence has not been found for either the tau-neutrino or the t quark. These are loose ends of the Standard Model that must be tied up at some point. If they are not found soon, we may be forced to begin classifying their nonobservation as an anomaly.

Generations of Quarks and Leptons

We group the quarks and leptons into three "generations," each containing two quarks and two leptons. The first generation contains the u and d quarks, the electron, and the electron-neutrino that compose the bulk of the ordinary matter of the universe, the substance of everyday experience. The second generation contains the c and s quarks, the muon, and the muon-neutrino. The third generation contains the t and b quarks, the tauon, and the tau-neutrino. Experiments performed as recently as late 1989 indicate that perhaps only these three generations exist.

These higher generations generally lead to the unstable particle states mentioned earlier, and so exist only momentarily in high energy collision processes. However, for a brief instant in the extremely hot relativistic gas existing during the first tiny fraction of a second after the universe began, the second and third generation particles were produced as copiously as first generation particles. So they must be considered equally important in any accurate description of the events ten to twenty billion years ago that resulted in the universe we observe today.

Antimatter

In an important indication of the basic symmetry of matter, each of the three generations of fundamental particles also exists in the form of "antimatter." Paranormalists often protest that scientists are too reactionary, too unconventional, too closed-minded to accept the possibility of other components of the universe, or other forms of reality, like spirit. The discovery of antimatter belies that contention, illustrating that scientists are perfectly willing to recognize the existence of a form of reality dramatically different from conventional knowledge, once adequate evidence for that component is demonstrated.

The small antimatter component of today's universe is rarely seen, except in high energy collisons in space, or by means of large particle accelerators on earth. However, in the early stages of the Big Bang, equal amounts of matter and antimatter most likely existed.

The 10^{87} photons that now pervade all space are the result of the annihilation of most of the matter and almost all of the antimatter of the early universe. Only one particle in a billion survived that holocaust to form the galaxies. If this tiny asymmetry had not existed, neither would we, and the universe would be nothing but microwave photons, neutrinos, and perhaps whatever composes the dark matter. And none of it would be visible.

Forces

To complete the Standard Model, another equally important component must be added: the forces by means of which the fundamental particles interact. Understanding the modern conception of how the forces between bodies are generated will be important when we start to talk about various paranormal ideas. We will see that these ideas are based on an outmoded, nineteenth-century view of continuous fields that has been superseded by a picture that involves only discrete bodies interacting at localized points in space. In the modern picture, both matter and forces are particulate.

Until recently, just four forces were known to exist: (1) gravity, which holds the planets, stars, and galaxies together; (2) electromagnetism, which holds molecules and atoms together; (3) the strong nuclear force, which holds the nuclei of atoms together; (4) the weak nuclear force, which causes some nuclei to be radioactive and provides the primary source of energy in the sun and other stars.

Exchanging Photons

In the 1950s, the electromagnetic force between two electrons was successfully reinterpreted as resulting from the exchange of photons, the particles or quanta of light. For example, the repulsion between two electrons occurs when one electron emits a photon and recoils backward, like the "kick" experienced when firing a gun. The second electron absorbs the photon and is repelled in the process, like a tin can hit by a bullet. This interaction is elegantly described by the Feynman diagram, shown in Figure 2.1. Note how the paths of the particles diverge, as we expect for a repulsive force.

Attractive forces, which exist between two electric charges of oppposite sign, cannot be explained so succinctly. Suffice it to say that physicists have successfully described both attractive and repulsive electromagnetic forces between charged particles in terms of the exchange of photons that carry energy and other physical quantities across the space between the particles.

Since the time of Newton, the problem of how bodies can interact with one another over great distances has been the subject of almost continual debate. In order to explain gravity and electromagnetism, nineteenth-century physicists invented the concept of invisible fields pervading all of space. These fields provided great computational utility, but they were never directly observed, making their reality questionable. Nevertheless, the field concept became firmly entrenched as the means by which invisible forces of every imagined property are applied, from the "animal magnetism" of Friedrich Anton Mesmer in the late eighteenth century to the "morphogenetic fields" of Rupert Sheldrake today.

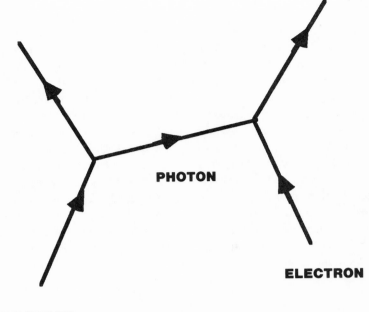

Figure 2.1. The Feynman diagram. The electron on the left emits a photon, recoiling backward. The electron on the right absorbs the photon, recoiling under its impact. Thus the repulsive force between two electrons is understood in terms of local processes. No action at a distance is necessary. All known forces can be understood in terms of similar particle exchange mechanisms.

Today we still maintain the language of fields for convenience in calculations, but the twentieth-century concept of particle exchange has made it possible to explain forces without the invocation of invisible, undetected fields.

Electroweak Unification

For almost thirty years after the success of the photon exchange picture of electromagnetic forces, physicists searched for analogous particles whose exchange might account for the strong and weak nuclear forces. However, success was not achieved until the quark-lepton structure of matter was clarified. The key to this development turned out to be the application of a principle used in electromagnetic theory called "gauge invariance." In the 1970s, Sheldon Glashow, Steven Weinberg, and Abdus Salam recognized that the gauge principle provided a mechanism by which electromagnetism and the weak nuclear force can be combined into a more basic "electroweak" force.

The idea that the electromagnetic and weak forces are equivalent requires an immense effort of imagination, since they appear vastly different on the scale of everyday phenomena. Even the highly creative thinker Richard Feynman found the proposal hard to believe. The problem is this: the electromagnetic force acts over cosmic distances. For example, light reaches us from galaxies across the universe. The weak nuclear force, on the other hand, falls to zero outside the range of the nucleus of an atom, 10^{-13} centimeters. How is it possible that these two forces are the same when their ranges of effect are so disparate?

The electroweak theory explained why in the everyday world we see electromagnetism and the weak force as very different phenomena: the structural differences we observe are not inherent and disappear at ultra high energies. They are like the differences in the patterns of two snowflakes that disappear when the temperature rises above freezing and they melt into identical drops of water.

As we raise the temperature, or energy, of a system, higher levels of symmetry can occur. Objects with different structures but the same underlying properties are seen to be inherently alike. In an analogous way, particles with similar properties but different masses become indistinguishable as the collision energies of the particles become far greater than their rest energies. And forces that are distinguished by their differing range at the low energies of normal experience (like the weak and electromagnetic forces) become equivalent at energies far higher than those of normal experience. The differences we observe are not inherent, just the result of the way the structures happen to freeze into place at low energies.

This concept of structure formation is called "spontaneous symmetry breaking." We can extend it to material structures of every type. As we will see, when particles collect into huge assemblages, they can spontaneously develop the properties we associate with life and mind. This happens by self-organization that occurs without the intervention of a supernatural designer.

Gauge Bosons

The theory of Glashow, Weinberg, and Salam implied that the particles exchanged in producing the weak force were the fairly massive objects called gauge bosons. This idea had been proposed years earlier. However, previous theories gave no idea of the specific value of the mass of the gauge boson. The new unification scheme made a major improvement. It predicted that the boson mass would be about that of a silver atom.

To test this prediction, two independent experiments, each with over a hundred scientific collaborators, were launched at the European Laboratory for Particle Physics (CERN) in Geneva, Switzerland. In 1983 each experimental group was able to report observations of the gauge bosons, called W and Z, at exactly the masses predicted by the gauge theory. Carlo Rubbia and Simon van der Meer received the 1984 Nobel Prize for their leadership roles in developing the techniques that were used.

The prediction and empirical verification of the W and Z bosons is a classic example of the scientific process in action: a theory based on fundamental principles and earlier observations allows a quantitative prediction that can then be tested by an experiment that is clear and unequivocal. If the prediction fails to be verified, the theory is discarded. How different from theories of the paranormal, which make no predictions that can be tested and are never discarded!

Nuclear Glue

While the unified theory of weak and electromagnetic forces was being developed, a parallel effort to explain the strong nuclear force along similar lines was being made by other theorists. Here again gauge bosons were introduced, as the particles exchanged in generating the force that holds quarks together in the proton and neutron. These gauge bosons were called "gluons," since they were the glue of atomic nuclei. A total of eight different varieties, or "colors," of gluons was found to be needed.

This theory of the strong nuclear force, dubbed "quantum chromo-dynamics," or QCD, by Feynman, has met with moderate success for about a decade now. Most QCD calculations, however, require the use of arbitrary parameters, so tests of the theory have not been completely satisfactory. Still, no one has suggested anything better, and QCD fits nicely within the framework of gauge theories of the type that worked so well for electroweak unification.

Gravity

Although electromagnetism and the two nuclear forces are described by the exchange of gauge bosons in the Standard Model, gravity has yet to be incorporated in that scheme. Einstein's theory of general relativity, published in 1916, describes gravity as an effect of the geometry of space and time. Remarkably, it remains to this day consistent with everything known about gravity, and so rightfully deserves a place in an extended Standard Model. But the Einstein picture of gravity bears little resemblance to the quantum gauge theories that describe the remaining forces.

General relativity preceded the development of quantum mechanics, and a quantum theory of gravity has yet to be developed. Physicists hypothesize a gauge boson, called the "graviton," as the particle exhanged in the gravitational force. Like the photon, the graviton would be exhanged between bodies separated by billions of light years away.

However, no evidence for the graviton, or any quantum theory of gravity effect, has yet been found. This does not mean gravity violates quantum mechanics, but merely that the weakness and long-distance nature of gravity result in such small quantum effects that they have so far defied observation.

Since general relativity is not a quantum theory, it must be regarded as provisional. Although it has worked beautifully so far, general relativity is incomplete, and we may anticipate the discovery some day of an experimental result that it cannot explain. Quantum effects must play a role at some level. Ultimately, a quantum theory of gravity will have to be developed. But until quantum gravitational effects are observed, we cannot rightfully say that any anomaly in the Standard Model exists on the basis of our current level of ignorance. Human ignorance is certainly no anomaly.

Summary of the Standard Model

As outlined above, the Standard Model contains three generations of quarks and leptons, and their antiparticles, of which only the first generation of particles constitutes the predominant matter of the visible universe. To these are added the gauge bosons exchanged to produce the forces between quarks and leptons: the photon, W and Z bosons, and gluons. All the known matter in the universe consists of these particles, although we have yet to determine the nature of the dark matter.

The forces of the Standard Model do not require the instantaneous action of invisible fields over separated regions of space. Gravity remains observationally consistent with general relativity, but quantum effects are expected to exist at some level.

Grand Unification

After the great success of electroweak unification and the more modest success of QCD, the next logical step for particle theorists was to attempt further unification of the remaining forces: gravity and the strong nuclear force that holds nuclei together.

The first idea, pursued with some theoretical success, was the unification of the electroweak force with the strong force in what is called "Grand Unification Theories" (GUT). The simplest specific GUT has the less colorful name "minimal SU(5)."

Like the Glashow-Weinberg-Salam electroweak theory, minimal SU(5) made a specific quantitative prediction that could be tested using existing technology. Theorists calculated that the proton should be unstable, living, on average, 10^{31} years before disintegrating. However, several large experiments that were set up to test this prediction failed to find proton decay. Minimal SU(5) failed to be verified.

Scientists must bow to the verdict of experiment. Although achieving the unification of three of the four forces of nature within the period of a single decade would have ranked with the highest intellectual feats in history, we had to accept what the data were telling us. So minimal SU(5) was discarded, although some theorists maintain that it might someday be resurrected with appropriate changes to its parameters.

Beyond the Standard Model

And so, the Standard Model remains at this date a non-unified theory, with gravity and the strong nuclear force still to be coupled to the electroweak force. This suggests that the story of the elementary objects and forces of nature has not reached its final page with the current Standard Model. We have a way to go before we can announce an ultimate "theory of everything."

Theorists will almost certainly continue to need the guidance of experiment to narrow the range of possibilities for the next edition of the Standard Model. This guidance will be the most useful when the experiments uncover anomalous results—results that cannot be explained by the Standard Model. And so, a new generation of high energy accelerators and other powerful instruments will soon be seeking out anomalies that it is hoped will point the way beyond the Standard Model.

Searching for Anomalies

Although experimentalists at existing laboratories currently have no anomalous data, they continually report "suggestions" of effects that "need further study." Perhaps some of these effects hint at new phenomena; however, most will likely not be confirmed. This is what usually happens.

The Superconducting Supercollider (SSC), the Hubble Space Telescope, and numerous other, less expensive instruments are now being prepared to seek out the next set of anomalies. I will be both surprised and very disappointed if none is found with these marvelous new tools.

The SSC, to be built in Texas, is the most ambitious scientific apparatus ever planned. An underground tunnel fifty-three miles long will contain two oppositely circulating beams of protons, each with an energy of twenty trillion electron-volts. The beams will be brought together and collided head-on, producing vast numbers of particles of every variety known and perhaps unknown.

The Standard Model has features that cannot still apply at the forty trillion electron-volt level of energy of the colliding proton beams of the SSC. The theory must break down, and the form that the resulting experimental anomalies take should point the way to the next Standard Model of fundamental constituents and forces of nature.

But when new anomalies are found, they are not likely to be classified as supernatural or paranormal. Few scientists will suggest explanations that go beyond matter. Rather, the anomalies will be assumed to be perfectly natural, and we all will get happily to work proposing new experiments and developing new theories to enable us to grasp what the new data suggest about the next layer of structure of the universe.

The energy obtained in the colliding beams of the SSC, corresponding to a temperature as high as 10^{18} degrees, existed naturally when the universe was only 10^{-15} second old. Thus, the physical processes observed in these collisions will approach those of another event that is usually regarded as supernatural: the origin of the universe.

The Early Universe

The universe we live in is very cold, -270° C, three degrees above absolute zero, except in the vicinity of energy sources like the sun or a nearby fireplace. In the first fraction of a second after its birth, however, the universe was many orders of magnitude hotter than the center of the sun. At an early enough time, the energies were higher than we can ever dream of producing in accelerators on earth.

Thus, the early universe serves as a laboratory for ultra high energy particle physics, and a few prominent physicists think that clues to the next level of unification could best be sought in the study of the origin of the universe. Even if that should not turn out to be the case, our understanding of the universe's beginnings has been dramatically advanced by the development of the Standard Model, and the insight this has provided into the possible natural, spontaneous origins of physical laws.

Out of Chaos

As we go farther back in time into the heart of the Big Bang, to higher and higher levels of energy, we approach a regime in which the universe is simpler and more symmetric than it is today. In the extreme of a completely symmetric universe, no structure of either matter or forces would exist.

Extrapolating back from current knowledge, we find that the universe would have been in a state of complete chaos at 10^{-43} second after the beginning of time. At that particular time, called the "Planck time," the universe was a black hole 10^{-33} centimeter in size. No structure could have existed, no particles, forces, or grand design.

At the Planck time, the universe just appears—as a quantum fluctuation. From what? We can't even ask, because we can't define an earlier time. And even if we could, if design and order existed in that time interval, they necessarily would be completely destroyed at the Planck time.

Whatever order now exists in the universe, including the laws of physics, developed after the Planck time. Even if these resulted from the act of a creator, they happened not at time zero but after the Planck time. However, a creator is unnecessary. We can now imagine how the particles of matter and the force laws they obey were generated naturally and spontaneously by a series of "phase transitions" as the universe exploded out of the void (Stenger 1988).

Snowflakes and Spheres

Phase transitions are familiar in everyday experience. Gases liquify, liquids vaporize or freeze, and solids melt. When a material undergoes a phase transition, two of its features undergo abrupt change. One of these is its level of symmetry, the other is its level of organization.

Let's consider the example of the water droplet and the snowflake. The snowflake has a more complex structure than the droplet, and so we say the snowflake has a higher level of organization. On the other hand, a snowflake is less symmetrical than a spherical droplet.

A sphere can be rotated about any of an infinite number of axes and will still look the same; it has what we call a very high level of symmetry. The snowflake is symmetrical, to be sure, but this symmetry is around only one axis and is restricted to rotations in steps of 60° about that axis. So the snowflake, with its six-fold symmetry, is still less symmetrical than the water droplet.

Parenthetically, note that our concept of beauty is not correlated to the level of symmetry. In fact, quite the contrary. Objects with interesting structure, such as snowflakes, are considered more beautiful than highly symmetric geometrical objects such as spheres. Yet the spherical droplet is considered more beautiful than the cloud of vapor out of which it condensed. And, when gravity breaks the spherical symmetry to produce a raindrop with its more familiar, pointed shape, the result is considered prettier than a sphere, but not so pretty as a snowflake.

By analogy, the high level of organization we see in life on earth is beautiful because of its structure. And that structure is the result of the breaking of the symmetries of an earlier, less-organized but more-symmetrical state.

The Origin of the Universe

Now let me try to show the connection between snowflakes and the structure of the universe. The snowflake is like the highly structured universe we observe today. This structure, we now believe, could have come about spontaneously by an analogous mechanism to the formation of a snowflake from water vapor.

The account of the origin of the universe according to this mechanism violates no known principles of physics, and requires no miraculous creation. The universe spontaneously began in complete chaos, with zero order and a high level of symmetry. No laws of nature existed other than those conservation principles implied by the high level of symmetry.

Except for these symmetries, no structure existed—all was randomness. Random quantum fluctuations, not initiated by any causal process, then led to the appearance of a tiny region of vacuum with maximum disorder and maximum symmetry. Why did this happen?

Why not? Given all possibilities, why shouldn't it have happened? And why not all other possibilities as well? Our universe was formed in one of the infinite number of ways it could have formed. The particular structure of our universe came about by chance, freezing into form just like the six points of a snowflake.

Einstein has shown that even an empty vacuum can have a geometrical space-time curvature with a uniform energy density and negative pressure. This would result in an exponential expansion of the space and rapid increase in the total energy contained in the curvature. Although Einstein himself made no use of this idea of exponential expansion, in recent years the idea has been resurrected in what is called the "inflationary universe" (Kasanas 1980, Guth 1981, Linde 1982, Guth 1984).

We are accustomed to an expanding gas, such as that inside the cylinder of an auto engine, producing positive pressure and losing internal energy as it expands, unless heat is added from outside. The heat, or more precisely the change in internal energy, drives our automobile engines as the hot gas presses against the pistons.

In the case of the expanding vacuum universe, however, we have a negative pressure. The vacuum then actually does work on itself, raising its own energy. Bizarre as this seems, it follows from general relativity and the curvature of space-time, and violates no principles of physics.

As the universe expands and cools, it passes through a series of phase transitions, and the energy released during each transition creates particles. The laws that these particles obey appear as increased structural order, like the pattern of a snowflake. When the previously existing high level of symmetry spontaneously breaks, a lower level of symmetry ensues.

This whole process occurred in a tiny fraction of a second at the beginning of the universe. Once the quarks and leptons were generated, forces spontaneously came into being that held the quarks and leptons together as matter. When, after a minute or so, the universe had cooled to the point that the composite nuclei and atoms that were produced were not ripped apart, the galaxies, stars, and planets could form (Weinberg 1977).

Eventually, on at least one planet revolving around a star in a universe of 10^{22} stars or thereabouts, atoms would further condense into complex molecules that accidentally acquired the ability to reproduce themselves. When this happened, evolution by natural selection took over, and particular structures that enhanced future reproducibility rapidly developed.

And so it happened, in a few billion years, without cause or plan and with no externally imposed fundamental laws of nature beyond those that accidentally evolved during the first few seconds of the inflating universe, that the high levels of material organization we call life and mind came to the planet we call earth.

References

Georgi, H. April 1981. "A Unified Theory of Elementary Particles and Forces." *Scientific American* 244-4:48.

Guth, A. H. 1981. *Physical Review* D23:347.

Guth, A. H., and Steinhardt, P. J. May 1984. "The Inflationary Universe." *Scientific American* 250-5:116.

Kasanas, D. 1980. *Astrophysical Journal* 241: L59-L63.

Linde, A. D. 1982. *Physics Letters* 108B:389.

Stenger, Victor J. 1988. *Not by Design: The Origin of the Universe.* Buffalo, N.Y.: Prometheus Books.

Weinberg, S. 1977. *The First Three Minutes.* New York: Basic Books.

3.

Doing the Devil's Job

Truth emerges more readily from error than from confusion.

Francis Bacon

The Devil's Advocate

One fine Wisconsin summer day in 1980 I stood before 500 fellow particle physicists at a major international conference. Representing a large research collaboration, I was given five minutes to present our evidence for gluon radiation in neutrino interactions. I didn't convince everyone. While I was conversing outside the hall after the session, a veteran experimentalist named Rod Cool (now deceased) came up to me and said, "Stenger, you had no right to make the claims you did in that talk!"

Thank goodness I was eventually vindicated when the paper based on the results was published in *Physical Review Letters,* the journal with the toughest referee standards in physics. Furthermore, subsequent observations in independent experiments confirmed the validity of the claims I had made on behalf of the collaboration.

So was my tormenter wrong? Should Professor Cool not have been more courteous? Should he not have accepted the fact that surely I knew more about this particular subject than he, and just taken my word as a gentleman that the evidence was there?

No. Much as I didn't like it at the time, he was quite right to confront me. Professor Cool did nothing worse than what I and other scientists have done many times: played the devil's advocate. As it turned out, his not-so-gentle pressure forced my collaborators and me to do a better job in the published paper. For this we thank him.

The role of the devil's advocate is fundamental to scientific method, and criticism by colleagues, unpleasant as it sometimes is, forces us to make the best possible case in presenting new results. When, as often happens, the evidence does not stand up under intense scrutiny, something useful still has been accomplished. We turn our attention away from unprofitable avenues of investigation, cut our losses, and move on to something else.

When we explore uncharted territory at the limits of knowledge, we have little guidance; so we are forced to explore in many directions. Most turn out to be blind alleys. Failure is more common than success. Being human and sometimes unwilling to admit defeat, we tend to keep at something long after any benefit can be expected from it. An outside critic, unburdened by any emotional attachment to the issue, is often needed to get us back on track.

A Rude Awakening

After years as a research scientist, I naively believed that everyone involved in the process of scientific discovery understood the necessary role of the devil's advocate in that process. Then I had a rude awakening. When I began applying to claims of psychic phenomena the type of hard-nosed, show-me criticism I was accustomed to both giving and taking as a particle physicist, I suddenly found myself subjected to personal attack. I began to see my name in print, not just as another coauthor of a scientific article, but denounced as a "witch-hunter," "McCarthyite," and even as someone possessed by the Devil!

Sued by a Psychic

My adventure began in the fall of 1986. I discovered that a psychic had been teaching noncredit courses on several campuses of the University of Hawaii. The courses had titles such as "Mastering Clairvoyance" and "Developing Telepathy." The course descriptions contained a number of incorrect and misleading statements, such as, "telepathy is a psychic skill documented around the world." In these course descriptions and other public pronouncements on radio, television, and in the press, the psychic promised that he would aid his students in developing their psychic skills, to enable them to defend themselves against "auric attacks" by "invading minds," and to "stand protected in haunted places."

Taking Action

As a scientist, I have a professional responsibility to help the public distinguish between fact and fiction. As a professor, I also share responsibility for what is presented in my university's classrooms. The public has a right to expect the highest quality of scholarship in its tax-supported institutions. So I thought I was merely doing my job when I wrote to the provost of the campus where the psychic was most active. In my letter, I rather strongly stated my opinion that the courses, as advertised, were not appropriate for a university classroom.

I did not contend then, nor do I now, that psychics, parapsychologists—or for that matter, anyone else—are not fully entitled to their personal beliefs or should be forbidden from expressing them freely on campus. Rather, I argued that this particular lecturer's published claims were not supported by the best available evidence. A university is obligated to present established knowledge in its classrooms—not an instructor's or a student's wishful fantasies.

With his wife as his attorney, he proceeded to sue me, the University of Hawaii, and several others, claiming that we had defamed him and deprived him of his livelihood and First Amendment rights.

Rushing to the Psychic's Side

Without trying to determine our version of the story or independently investigating the facts, the national occult community and a number of prominent parapsychologists from around the country rushed to the psychic's side. In their national journals and magazines, and in letters written to the president of the University of Hawaii and the local media, they deplored what they incorrectly viewed as an attack on parapsychology.

Actually, none of the defendants in the lawsuit had ever objected to the teaching of parapsychology, and if the protesting parapsychologists had taken the trouble to check the facts they would have learned that the plaintiff had made public statements about parapsychology that violated the field's own published principles. For example, the "Methods of Parapsychological Research" of the American Parapsychological Association clearly state that caution should be exercised in claims that one can be "trained" in psychic ability. If parapsychology is ever to gain credibility as a science, parapsychologists will have to abide by the hard-nosed critical methods of science.

Complete Vindication

On August 23, 1988, the lawsuit was dismissed and the defendants vindicated on every count. In his ruling, Judge Harold Fong of the U.S. District Court declared that the lawsuit was "frivolous" and suggested that it was brought by the plaintiff "to chill defendants' exercise of their rights rather than to vindicate his own rights." As for the fundamental issue of academic freedom, the judge said that "the classroom is not a public forum." A university is under no obligation to provide a classroom for anyone who comes in off the street to teach a class on whatever he or she wishes. This is no violation of free speech. Most campuses provide open forums where any and all ideas can be freely expounded.

My biggest surprise in this whole affair was not the reaction of the psychic's supporters from the occult community, which was to be expected, but the unquestioning support he received from the national parapsychological community, including several academics. Up to that time, I had believed and stated publicly that parapsychology was a legitimate scientific discipline, and that its practitioners held to the same scientific principles and had the same dedication to the truth as scientists in every other recognized discipline. At least they claim such in their published writings.

Pseudoscientists and Pseudoskeptics

When chastened by Rod Cool, my reaction was to assemble my collaborators so we could examine our data more closely, make other careful checks, and try to directly answer specific criticisms. It wouldn't have occurred to us to respond by attacking Professor Cool, questioning his motives or trying to find some skeleton in his closet the exposure of which would reduce his credibility and thereby raise ours.

In our understanding of devil's advocacy, the issue was the validity of our claims. Those claims would not be made any more valid if we had sued Rod Cool for libel. If a parade of Nobel laureates had testified to our outstanding personal qualities and utter trustworthiness, and a jury had ruled in our favor, our physics results still could have been wrong. I can honestly report that we were sincerely interested in the truth, and if upon reevaluation our reported effect had disappeared, we would have withdrawn the paper and moved on to something new.

Unfortunately, on the fringes of science we find a few people who either do not understand the value of critical thinking, or actually undermine it for purposes other than the quest for truth. They assume the dignified cloak of science, paying lip-service to its principles, but refusing to be bound by them. These are not scientists, but pseudoscientists. They think they can have it both ways—labeling what they do as science, while being free from the strict rules of evidence that conventional scientists are forced to obey.

When stung by a rising chorus of criticisms of their methods, pseudo-scientists have fought back recently. They label as "pseudoskeptics" everyone who may be less compromising, less willing to accept the shoddy data, sloppy methodology, and downright dishonesty that is often passed off on the unsuspecting public as science. The pseudoscientists go on to say that the "true skeptic" will examine both sides of an issue and be equally skeptical of the arguments against the existence of some abnormal phenomenon as of the arguments for it. That sounds fair, doesn't it? But it doesn't work that way.

Sophistry

The ancient Greeks had a name for the notion that all ideas and opinions are equal. They called it "sophistry." In Plato's *Theaetetus,* a young Athenian tells Socrates, "The sophists claim that everything is true according to each individual's measure of truth, and thus all theories are equally true and false." Socrates astutely replies, "Then I would say that they must admit that their own statement can be false too!"

Must a skeptic be skeptical of everything in order to be true to the term? I claim not. I don't have to be skeptical of the existence of electrons or the validity of established principles such as conservation of energy. I can see the light go on in the room when I flip the switch to start the current of electrons flowing, and understand that I must regularly stop by the filling station to keep my car going. Why must I be skeptical of the vast body of knowledge accumulated in science and other disciplines over the centuries? When something clearly works over and over again, without exception, why should I have to doubt it or give up my right to call myself a skeptic?

On the other hand, I feel great responsibility as a professional scientist to be skeptical about claims that are in violation of what experience and science has taught us about the world. I am not convinced that everything is possible, if only you "try hard enough." Flap your arms as hard as you will, you are not going to fly. I know for sure that I will never be a great poet or pianist. I know for sure that the most talented ballplayer of all time won't be able to get 1,000 hits in 1,000 times at bat. And I know for pretty sure that ten thousand or ten million people simultaneously thinking "lift" will not cause a microgram of matter to lift a micrometer off the ground.

Pragmatic Skepticism

In his book *A Physicist's Guide to Skepticism,* Milton A. Rothman distinguishes between ideological skepticism and pragmatic skepticism. The ideological skeptic has a deep-seated psychological disbelief in everything because of a dislike for all authority. This is clearly an unhealthy frame of mind. The pragmatic skeptic, on the other hand, disbelieves phenomena that contradict the principles or laws of nature that have been thoroughly corroborated by experiment (Rothman 1988).

I am recommending pragmatic skepticism here, a healthy and positive, if not always "kinder and gentler," force in human affairs that is fundamental to the evolution of ideas. I claim I can be constructively skeptical of holistic medicine while awaiting the evidence. And yet, because of the statistical record, I can believe that smoking causes cancer and that fluoridation prevents tooth decay.

We skeptics are free to believe something when the evidence is strong, as long as we are willing to change our views when the evidence swings the other way. Put another way, we can *almost* be dogmatic. We should always leave the window open a crack for the admission of a worthy new idea, although not quite so wide as to allow all the rubbish flying around outside to be blown in.

Good Theories and Bad

I have heard somewhere that the great physicist Wolfgang Pauli once made a scathing remark about a theory that had been ineptly presented to him. "Why," he reportedly said, "this theory is so bad it's not even wrong!" What he meant, I think, was that a good theory offers a way to be tested— usually in the form of a highly specific, unique, and preferably quantitative prediction, whose lack of confirmation will kill the theory. A good theory should provide the means of its own potential destruction. A theory that does not show us how to destroy itself is "not even wrong."

The fact that a theory may eventually test wrong does not detract from its original merit as a worthy try. On the other hand, if an idea is poorly formulated, often because the terms used are not clearly defined, then how can we even test it? When confronted with an irrefutable proposition, skeptics are under no constraint to spend time and trouble to find refuting evidence not already on the table. We are not pseudoskeptics when we reject out of hand something that is not testable or open to refutation, not because it disagrees with our own views, but because it provides no means of being judged right or wrong. We cannot determine that gibberish is anything but gibberish.

Testability is a requirement of scientific method that is agreed upon by all participants in the process because, without it, no reasonable debate is even possible. The successful track record of the testability criterion testifies to its usefulness—indeed its power—in settling scientific disputes. Someone who rejects the notion that all theories must be subject to test rejects scientific method.

Theories That Explain Everything

Early in this century, the philosopher of science Karl Popper grappled with the problem of distinguishing science from pseudoscience. He realized that science is not simply a matter of explaining already well-established empirical observations after the fact; pseudosciences, such as astrology, do that. He proceeded to compare four revolutionary developments that were in progress at that time, all labeled "scientific" by their promoters. These were: (1) Einstein's general theory of relativity; (2) Marx's theory of history; (3) Freud's psychoanalysis; and (4) Alfred Adler's "individual psychology."

Popper tells how excited he became when, in 1919, Arthur Eddington confirmed Einstein's prediction that light from a star would be bent in passing by the sun. He compared this specific, quantitative result with the qualitative and untestable claims being made in the other three areas and concluded: "I felt that these other three theories, though posing as sciences, had in fact more in common with primitive myths than with science; that they resembled astrology rather than astronomy" (Popper 1987).

What made the difference in Popper's mind? In a sense, these theories were too explanatory. Popper observed that, in the minds of the supporters of Marx, Adler, and Freud, the three theories explained everything within their fields. They had the effect of an "intellectual conversion or revelation, opening your eyes to a new truth hidden from those not yet initiated." The truth of these theories was self-evident to the converted. They were more religion than science. Far from providing no evidence, verification was found everywhere one looked.

Popper gave a perhaps oversimplified illustration of how Adler's and Freud's theories could be used to give incompatible explanations of human behavior. The illustration uses alternate responses to a situation as examples. In the first example, a man tries to drown a child by pushing him into the water. Freud would say that the man suffered from some repression, such as an Oedipus complex, while Adler would say he had an inferiority complex and needed to prove himself.

In the second example, another man saves a drowning child. Freud and Adler have no problem explaining this either. According to Popper, Freud would say that the man achieved "sublimation," while Adler would say that the man had an inferiority complex and rescued the child to prove himself.

Thus, Freud and Adler each have a theory that explains each outcome equally well. No matter what happens, whether the child is drowned or saved, both theories are verified. No test is provided by either theory to enable us to judge whether it is correct, or indeed to distinguish it from the other theory.

Popper's criticism of the two theories focuses on their overly broad explanatory power. But that is not the only problem. Note that neither Freud nor Adler tells how to take their theories and apply them to a given individual, predicting whether he will drown or rescue the child. By allowing both possible outcomes, their theories are empty of predictive power. They are ad hoc and ex post facto—with a ready explanation to pull out of the filing cabinet for whatever happens and only explaining after the fact.

Most scientists I know have little use for theories with limited predictive power. However, philosophers of science do not totally disdain after-the-fact explanations. Philosopher Larry Laudan has pointed out that one of the most important successes of Einstein's general theory of relativity was actually a "postdiction," its explanation of the precession of the perihelion of the planet Mercury.

This precession effect had been known for centuries, and no one had succeeded in explaining it within the classical Newtonian framework. Many scientists regarded Einstein's explanation of the Mercury precession at least as significant as his prediction that light would be bent by the sun.

However, the difference between Einstein's ex post facto explanation of the Mercury precession and the examples of Freud, Marx, and Adler is that the general theory of relativity succeeded where all other attempts failed. The theory also went on to make a range of surprising and novel predictions, such as the bending of light by the sun, that were triumphantly confirmed. Freud, Marx, and Adler succeeded in distinguishing themselves from their rivals for reasons other than the predictive power of their theories.

Falsifiability

Popper felt that the key ingredient making Einstein's theory good science was not the fact that his predictions were *verified*, but rather that they were capable of being *falsified*. That is, they could have been shown to be wrong and were not. These two concepts may seem equivalent, but there is another ingredient to falsifiability. Einstein's general theory of relativity made predictions that, had they failed to be confirmed, would have ruled out the theory. Thus, his predictions were also risky.

Popper recognized that no amount of verification will ever guarantee that a theory is valid and complete. As he put it, "Irrefutability is not a virtue of a theory (as people think) but a vice" (Popper 1959). A good theory makes specific and, most important, risky predictions that can be tested, not to prove its correctness but to prove its usefulness. It must be falsifiable.

Theories that explain everything explain nothing. Bad theories, as Pauli said, are not even wrong. And so, pseudoscience is not science because it is wrong; it is pseudoscience because it is useless. In science, a theory can turn out to be wrong. But it is better to be wrong than to be useless.

A Good Theory: Electroweak Unification

An example that both Popper and Pauli would probably have agreed is a good theory can be found in recent particle physics history. As described earlier, the electroweak unification scheme of Glashow, Weinberg, and Salam rested on the existence of previously undiscovered particles called gauge bosons (W and Z), predicted to be eighty and ninety times more massive, respectively, than the proton. This was a highly specific and risky prediction. When the W and Z particles were reported at the predicted masses by two separate experiments in Geneva in 1983, electroweak unification was upheld.

If the gauge bosons had not been found with the predicted properties, the theory of electroweak unification would have been discarded in the form proposed. Instead, it emerged triumphant. However, we should note that the nonfalsification of electroweak unification does not mean it is true. The opposite of false is true; but the opposite of falsification is not truth, but only the tentative possibility of truth.

At this point, some physicists and other scholars familiar with this particular history might object. Suppose the experiments had turned out other than they did and the bosons had not been found. Then theorists, who are famous for their flexibility, would have cooked up an explanation for the failure of the prediction, and continued to push for some form of unification. The result would be a new version of the theory with different parameters and with its own distinctive predictions. The earlier experimental result still would have been crucial in ruling out the previous set of parameters and in turning attention away from a popular but unprofitable line of investigation. Even if the W and Z bosons had not been found, electroweak unification would still have been a good theory, according to the criterion of Popper and Pauli.

What Makes a Theory Acceptable?

So falsifiability, the ability to be ruled out by experiment, is a necessary condition for any theory to be taken seriously. However, this is not a sufficient condition. A theory that predicts the world will end on Halloween night in the year 2000 is falsifiable. But before making any plans for the end of the world, a prudent person would demand additional criteria. What should these criteria be? Based on the previous examples, for a theory to be accepted it must not only make specific unique or risky predictions, but some of them must be confirmed.

A theory predicting the world's end is useless if it can only be tested on that single outcome. For such a prediction to be taken seriously, it should have some other tests we can apply before the final cataclysmic event. For example, the theory might predict a rise in ocean levels of a certain number of feet per year, or it might tell us to look in the sky on a certain date for an approaching comet from the direction of Orion. These are unlikely events that could be precisely checked.

The antithesis can now be drawn: What makes a theory bad? A bad theory is not necessarily one that eventually disagrees with observations. Bad theories are theories that make no risky predictions, that explain everything but nothing unique, that are basically useless.

Anything Goes

What I have espoused here is essentially the conventional view of scientific method, poorly understood though it may be outside of science. This consensus has been challenged by several influential contemporary philosophers, sociologists, and historians of science.

Thomas Kuhn, in his book *The Structure of Scientific Revolutions,* promotes a doctrine that science contains accepted bodies of knowledge, "paradigms," very difficult to change because of science's basic conservatism (Kuhn 1970). Supposedly, only a revolution, such as occurred with Darwin's theory of evolution, Einstein's relativity theory, or the theory of quantum mechanics, leads to major changes in the accepted paradigms—paradigm shifts. These changes are forced upon the majority of scientists, who must be dragged kicking and screaming into accepting the new paradigm.

The popular image of scientists in action imagines short bald men in white coats peering into microscopes away in the laboratory. They are absentminded, forgetful, and unconcerned about the outside world, selflessly dedicated in their pursuit of truth wherever it may lead.

But Kuhn and his followers paint a different picture. They see a scientific establishment forming a social unit not fundamentally different from other professional groups such as lawyers or dentists. Like these, the scientists are primarily motivated by self-interest and influenced by the parochial views of their colleagues. Science, then, is driven by social forces acting to maintain the reigning paradigm, rather than the drive for truth.

Kuhn has been strongly seconded by Paul Feyerabend, who went even further in claiming that the historical record of scientific inquiry does not exhibit the rational pursuit of knowledge. He argues that, in the overwhelming majority of major scientific shifts, scientists resisted the changes, even when these changes were strongly supported by factual evidence.

Scientific progress is possible, in this view, only when individuals such as Galileo break sharply with the accepted practice and refuse to follow the rules of their particular scientific epoch. They must behave as intellectual mavericks or anarchists, more in the popular image of Dr. Frankenstein. According to Feyerabend, progress in science crucially depends on such seemingly irrational behavior of its practitioners, and "the only principle that does not inhibit progress is: *anything goes*" (Feyerabend 1975).

Spoken like a true man of the sixties. I can just imagine Socrates' retort to this modern-day sophist: "If *anything goes,* then so does *nothing goes*" (as suggested by Theocharis and Psimopoulos 1987).

I must say that I have often seen and been personally dismayed by the effects of fashion in my own field of particle physics. People follow the lead of a small group of influential figures, often down blind alleys. However, the changes that occur as science progresses are simply not as abrupt as Kuhn and Feyerabend imply, with each important new discovery the result of some Lone Ranger fighting to gain acceptance of his radical ideas. Paradigms in fashion at any given time are not as inflexible as the neosophists would have us believe, buried in concrete that must be blasted away with terrorist dynamite. Change from one paradigm to another can and does most often occur gradually, systematically, in small steps.

The seeds of great new ideas, such as Einstein's relativity, can be found in earlier published work. As Larry Laudan has said about Kuhn's paradigm shifts, "This kind of tunnel vision, in which a sequence of gradual shifts is telescoped into one abrupt and mighty transformation, is a folly which every historian is taught to avoid" (Laudan 1984).

Furthermore, the Kuhn-Feyerabend thesis is refuted by the facts. They should have looked at modern-day physics. Nothing in the social structure of physics has prevented it from making startling new discoveries, such as gauge bosons and high-temperature superconductivity. No ultraconservative inbreeding has prevented physicists from reaching out to the stars for neutrinos, or probing into the heart of the Big Bang.

The Star System

Feyerabend and Kuhn have not seen modern physics in action from very close up. The deliberate process engaged in by the majority of physicists, geniuses along with conventional footsloggers, has led to progress and revolutionary change at the deepest level.

In science, as in entertainment and politics, we have a "star system" in which certain individuals get the lion's share of public recognition for major accomplishments. So it is not surprising that outsiders who learn about science by reading newspaper and magazine stories get the impression that a great new discovery was only made possible by the inspiration of some individual, that it was a brilliant flash of insight he alone received that showed him what the rest of us were doing wrong. This rarely happens.

Furthermore, the stars of science, such as Feynman, Weinberg, and Rubbia, were not mavericks who came in and straightened the rest of us out. They were tenured professors at mainstream universities and laboratories. Galileo and Newton had the education and credentials of the establishment. People like to note that Einstein was a patent clerk in that stupendous year of productivity, 1905, when he proved the atomic nature of matter, invented relativity, and proposed the particle nature of light. But he was also finishing up his Ph.D. dissertation for the University of Zurich at that time, and soon had a more conventional academic position. I can't think of a single fundamental discovery in physics for centuries past that was made by a true outsider.

The Bozos

In articles in the magazines *Science* and *Nature,* Paul Feyerabend is quoted as saying that "normal science is a fairy tale . . . equal time should be given to competing avenues of knowledge such as astrology, acupuncture, and witchcraft" (Broad 1979, Theocharis and Psimopoulos 1987). The fact that science works and these other practices do not seems not to have impressed Professor Feyerabend. I am not surprised when the average layman confuses astronomy with astrology, but I would expect a philosopher of science to be aware of the enormous gap between the two. Studies have shown that astrology has no predictive power (see, for example, Hines 1988, p. 141 and references therein), while the enormous predictive power of astronomy has amply been demonstrated. With astronomy, we can predict an eclipse of the sun a thousand years in the future.

As you might expect, pseudoscientists are greatly comforted by the neosophist critics of science. They say that scientific closed-mindedness is what keeps paranormal paradigms from being given proper attention and credibility, and fancy themselves as modern Galileos who must do battle with the entrenched scientific establishment.

No one ever said science was easy, and nobody, scientist or not, should be expected to fall over and play dead when a challenge to existing knowledge is made. If a new idea has sufficient merit, it should ultimately overcome any resistance, no matter how strong. Galileo was censored by the most powerful institution in Europe, the Catholic Church, yet his discoveries spread almost instantaneously throughout Europe.

And even if we can find examples where some great idea was slow to be accepted, it does not follow that the nonacceptance of a nutty idea results from anything other than its nuttiness. As Carl Sagan has said: "They laughed at Columbus, they laughed at Fulton, they laughed at the Wright Brothers. But they also laughed at Bozo the Clown." The Newtons, Einsteins, and Fultons eventually rise to the top—while the Bozos continue to roll around in the sawdust of the circus floor.

Resistance to new ideas is part of the process of science. A worthy new idea must overcome barriers of doubt and skepticism, and even occasional irrational objections. But if an idea has merit, it will eventually climb over these barriers.

Science in Practice

I will readily agree that the ideal methods of the textbooks and the linear logic described in scientific papers do not faithfully describe the procedures that were actually followed by the authors when they performed their research. The stops and starts, backtracks and false leads, are not reported. Intuition, emotion, and even irrationality are part of the process, and mistakes are made.

Scientists are people after all, and their own individual interests occasionally outweigh the facts. While occasional cases of fraud in the conventional sciences may surface, it is infrequent (although, sadly, beginning to appear in medical research). Fraud is rare in science, not because scientists are more virtuous than other people, but simply because scientific cheaters rarely prosper. All reports of important new phenomena are independently checked many times over.

If faulty conclusions arise in conventional science, the cause is more often incompetence and unintentional misrepresentation than outright fraud. But more common still is self-delusion. Investigators often see what they have a subjective interest in seeing in the data, rather than what is really there. People are unreliable witnesses in either the law court or the laboratory. And scientists are no less immune to inaccurate and subjective judgment than the average person on the street.

N-Rays

A sad example of self-delusion can be found in the case of N-rays. In the early part of this century, the discovery of the various forms of natural radiation had brought great renown to Wilhelm Roentgen, J. J. Thomson, Henri Becquerel, Marie Curie, and others. Hence, the 1903 report by R. Blondlot of the University of Nancy in France of emanations from various metals just barely visible to the naked eye came as no surprise, and Blondlot's findings were quickly replicated by others. The French Academy published about a hundred papers on the subject and gave Blondlot a medal and a cash award.

However, not everyone who looked for the effect could confirm its presence. So an American scientist, R. W. Wood, visited Blondlot's laboratory to study his methods directly. Not seeing the N-rays that Blondlot claimed to see, Wood surreptitiously removed a prism that supposedly bent the N-ray from the apparatus. Not a very nice thing to do, to be sure, but the cause of truth was served. Blondlot continued to see his N-rays, suggesting that the French scientist had been deluding himself all along. (For more details, see Gardner 1957 and Hines 1988.)

Clearly, one of the flaws of Blondlot's experiment was the subjective nature of his observations. Today we prefer instruments that provide objective, quantitative results—dial readings, digitized voltages—to subjective nonquantitative human observations. Human judgment simply cannot always be trusted.

As we will see, one of the main problems with psychic research is its reliance on questionable human testimony. The N-rays case also illustrates why scientific method demands the consistent replication of observations by many independent observers. A phenomenon is not accepted as scientific fact until its observation becomes almost commonplace. Reports of the paranormal always fail to meet precisely these criteria.

The Fusion Delusion

As we will repeatedly see, self-delusion is a common human trait that has contributed to much of the bad research that gets published in every field, not just parapsychology. That physical scientists are no more protected against self-delusion than anyone else has been amply illustrated recently in a celebrated case.

Just before Easter 1989, two chemists, B. Stanley Pons from the University of Utah, and Martin Fleischmann of the University of Southampton in England, announced that they had produced nuclear fusion in a simple, inexpensive experiment at the Utah lab. The chemists claimed to have achieved fusion at near room temperature by simply squeezing deuterium (heavy hydrogen) nuclei together within the crystal lattice of the metal palladium (Fleischmann and Pons 1989).

Pons and Fleischmann said they were able to generate considerable energy in their reaction, a remarkable and unexpected result. If true, it could solve the world's energy problems for the indefinite future. However, they initially lacked the instrumentation needed to convincingly demonstrate that they had in fact observed a nuclear reaction.

Attempts to Replicate

In the weeks that followed, research groups around the world attempted to repeat the experiment. Some claimed success, but many more said they had failed to detect any evidence for low temperature fusion in the manner outlined by Pons and Fleischmann, or they obtained inconsistent results.

In early May, a meeting of the American Physical Society heard some forty talks on both experiments and theory. The strong consensus of the speakers and audience was that the fusion process does not occur, at least not at the intense level reported by Pons and Fleischmann. Many of the latest experiments had been conducted at major research centers that had far more extensive instrumentation and other resources than were available in the small lab in Utah.

As of this writing, experiments in cold fusion are still being carried out in many laboratories around the world, including a new fusion research center in Utah. Some positive effects continue to be reported, but most of the experiments fail to confirm the reported phenomenon at anywhere near the level originally claimed.

The fact that a low temperature fusion process probably does occur at a level trillions of times lower than reported by the Utah team does not represent a confirmation of the Pons and Fleischmann observations. What they specifically claimed to see involved huge energy production. They were almost certainly mistaken. It appears that Pons and Fleischmann had fooled themselves into thinking they had stumbled upon the discovery of the century.

The scientific method ultimately triumphed. And it would have worked just as well and as quickly had the effect been real. Far from denying the chemists the right to state their claims, the scientific community gave Pons and Fleischmann every opportunity to do so. For a few months they had the attention of the world.

No one was prevented from studying the concept and testing it, if one had the know-how and tools. And, if the ultimate rejection of cold fusion comes about because of skepticism, skepticism did not make rejection mandatory. Nature, not skeptics, will decide the issue. And that's the lesson I wish to draw from the cold fusion incident, as it applies to the subject of this book. Nature, not human beings, will ultimately decide whether there is more to the world we inhabit than physical matter.

A Low Threshold

In the case of cold fusion, the news media reported scientific results before they appeared in the professional literature. This used to be rare, but happens more frequently these days. The danger is obvious: almost anything an investigator says can find its way into print someplace before it has been subjected to the critical analysis of experts. A good rule is to be wary of any report until it has run the gauntlet of the peer referee process and has appeared in a scholarly journal.

Even so, results are frequently reported in the professional literature of every scientific discipline that fail to be independently verified by later investigators. But this confirms, rather than negates, the self-correcting nature of the scientific process. The results that prevail are those that in the long run are supported by other workers. The publishing thresholds of scientific journals are not set so high that unconventional results will necessarily be rejected. In particle detectors, telescopes, and scientific journals, a little noise is accepted so that a significant signal is not missed.

Pathological Science

Chemist Irving Langmuir, who won the Nobel Prize in 1932, called examples of scientific self-delusion "pathological science." Recently a taped record of a lecture he gave on the subject in 1953 has been published (Langmuir 1989). In this lecture, Langmuir describes several examples of pathological science, including Blondlot's N-rays and the "Mitagenetic rays" that several scientists reported seeing in the 1920s. The latter were supposedly special emanations given off only by living organisms.

Langmuir reported on his own studies of UFOs. He found that most UFO reports were simply the planet Venus, seen in the middle of the day when people do not expect it. Years later, Jimmy Carter would be similarly fooled and order a study of UFOs when he became president.

Langmuir was unimpressed with the claims of extrasensory perception. He tells of a visit with parapsychologist J. B. Rhine at Duke University around 1934. Rhine told the skeptical chemist that he had millions of cases where results were obtained that could not be attributed to random chance. But when he probed further, Langmuir found that Rhine had excluded from his sample large amounts of data that had negative results, on the rationale that the particular subjects tested "did not like him" and so had deliberately tried to score low. Curiously, this data sample was also the one with the best controls.

Langmuir identifies the following "symptoms" of pathological science:

(1) The maximum effect that is observed is produced by a causative agent of barely detectable intensity, and the magnitude of the effect is substantially independent of the intensity of the cause.

(2) The effect is of a magnitude that remains close to the limit of detectability or, many measurements are necessary because of the very low statistical significance of the results.

(3) There are claims of great accuracy.

(4) Fantastic theories contrary to experience are suggested.

(5) The ratio of supporters to critics rises up to somewhere near 50 percent and then falls gradually to oblivion." (Langmuir 1989)

The Standards of Criticism

Paul Feyerabend says that the standards of scientific method are really "standards of *criticism:* rational discussion consists of an attempt to criticize, and not an attempt to prove or make probable" (Feyerabend 1978). Though Feyerabend was being critical himself, I'll buy this characterization. It says to me that criticism and skepticism are important components of the method by which we develop knowledge.

When I make what I think is a new discovery, my colleagues down the hall try to shoot me down, and usually succeed. When they come up with new schemes, I grab my Satanic spear and try to punch holes in their arguments. This is how the process works. And not just in science either. My academic colleagues in nonscientific disciplines behave the same boisterous way, paying no heed to their mothers' admonitions not to say anything unless it's nice.

Inspired by Aristotle, even the Catholic Church uses critical analysis, and has since its earliest days, as evidenced in the writings of the great theologians Augustine and Thomas Aquinas. In fact, the term "devil's advocate" refers to the Church-appointed investigator whose job is to try to find other explanations for the miracles attributed to individuals proposed for sainthood. The problem with the Catholic Church is not its logic, but its basic assumptions. Theology is the logic of theism, which rests on the proposition of the existence of God.

The Burden of Proof

Even in the case of a well-formulated and well-grounded proposition, the devil's advocate is under no obligation to go far out of the way to seek evidence to support his or her skepticism. I often find myself confronted with this basic misunderstanding of the critical method. The argument is made that I must do a large amount of research on my own and become familiar with all the literature on a subject before I am qualified to comment.

Now, of course, we certainly have no right to criticize from a framework of ignorance, but how much can the skeptic reasonably be expected to do? Scientific methodology demands that the burden of proof lies with the promoters of a new result, not with those whom they are trying to convince. The promoters must make their own case. In their papers and presentations, they must provide their own careful estimates of the various potential sources of error, and show clear evidence that these can be ruled out with high probability. They have the responsibility for selecting the three or four best reports on their findings and recommending these for critics to study.

Of course, to be able to judge these reports adequately, the critics should be familiar with the basic issues, the technology, and the rules of statistics. But they are under no obligation to dig out data that are not reported, though some have done so. And if the presented data do not justify the conclusions, then skeptics have every right to point this out and refuse to accept the conclusions. Why would someone hold back data unless that data failed to support his or her contentions? Data held back can be taken as prima facie evidence against the claims being made.

Guilty Until Proven Innocent

The public's sense of fairness and justice leads it to be too lenient in its judgment of extraordinary claims. Most people are willing to grant that others are honest unless proven otherwise. Those who take the opposite view are often regarded as boors. The layman has learned from TV and movie courtroom dramas that a person is assumed innocent of a crime until proven guilty. Scientific method, however, holds that the opposite should be done when evaluating a claim to a new discovery: in science, you are guilty until you prove yourself innocent!

When confronted with a claim of a new discovery, a plausible working hypothesis for the devil's advocate is that the result is wrong. The promoter must then prove his or her own innocence of the implied charge of incompetence or dishonesty. If this strikes laymen as unfair, they need to understand that science is the most successful method that the human mind has been able to invent for uncovering the truth, precisely because it takes this hard-nosed attitude.

Poltergeists and Other Strange Powers

I should not be surprised if the average person does not understand scientific method, but I really expected something better from Arthur C. Clarke, whose early books and stories had great influence on me as a youngster. In *Arthur C. Clarke's World of Strange Powers,* a book based on a television series in Britain, the following statement is made: "Many poltergeist cases cannot be easily explained away. . . . Only the supporters of the fraud hypothesis contemplate the problem with any degree of certainty, and they have not managed to produce evidence in many cases" (Fairley and Welfare 1984).

I strongly disagree with the implication that we should believe in the reality of a phenomenon simply because, for lack of information or whatever reason, no natural explanation has been made. Paranormal fraud does not have to be proven by the skeptics. It has to be disproven by the paranormalists.

In fact, skeptical investigators of poltergeists frequently find evidence for fraud. For example, in a case in Columbus, Ohio, in 1984, a fourteen-year-old girl who had received much media attention for her ability to call up "noisy spirits" was secretly photographed tossing a phone across the room while others in the room had their attention directed elsewhere (Randi 1985). There are many other cases where reasonable cases for trickery can be made or where fraud has been actually demonstrated. Thus, trickery remains a plausible working hypothesis for explaining poltergeists and most other reports of the paranormal.

Eyeless Vision and other Miracles

In the "Rationale" to *Supernature,* Lyall Watson illustrates how profoundly proponents of paranormal beliefs disagree with the methodological point of view I express here. He refers to an article in the March 1965 issue of *Scientific American,* "Eyeless Vision Unmasked," that tells how the Russian psychic Rosa Kuleshova was caught cheating. Watson remarks: "Several books since that time have used this report as justification for dismissing the entire phenomenon, but reference to the original research shows that, despite the fact that she was once caught cheating very clumsily at a public performance, Kuleshova also possesses a talent that cannot be reasonably shrugged off in this cavalier fashion" (Watson 1973).

As the philosopher David Hume said in his essay "On Miracles": "No testimony is sufficient to establish a miracle, unless the testimony be of such a kind that its falsehood would be more miraculous than the fact which it endeavors to establish" (Hume 1748). He gives this example: "When anyone tells me that he saw a dead man restored to life, I immediately consider with myself whether it be more probable that this person should either deceive or be deceived, or that the fact which he relates should really have happened."

Proponents of the paranormal like to counter this argument by pointing out the application of this rule by Thomas Jefferson: "Gentlemen, I would rather believe that those two Yankee professors would lie than to believe that stones fell from heaven." Of course, stones—meteorites—do fall from heaven. Jefferson should have had more confidence in Yankee professors. But this helps prove my case, not disprove it. Even the powerful objection of a president of the United States is insufficient to halt the progress of science, when that progress entails a phenomenon for which ample confirmatory evidence is ultimately found.

The Most Likely Hypothesis

Following Hume, skeptics like Thomas Jefferson accept the fraud hypothesis as the likeliest in certain cases, even when no evidence for fraud exists. How much more likely then is the fraud hypothesis when there is actual evidence for fraud, as in the Kuleshova case? Clearly, fraud need not be proven in all or even most cases. It need not even be proven in one case, just shown to be insufficiently ruled out, while alternative explanations were still possible.

One case of proven cheating on the part of a particular paranormal subject makes the cheating hypothesis more likely for any other claims involving that subject, in the absence of independent confirmation. And the large number of cases of spiritualists, mediums, psychics, and others having been caught cheating makes the cheating hypothesis one of the most likely explanations for their paranormal claims. Yet paranormalists ask us to discard virtually the whole edifice of science on the basis of a few obvious magic tricks of a Rosa Kuleshova or Uri Geller.

Science and the Supernatural

The experience of the distinguished British physicist John Taylor is particularly illuminating. In 1973 Taylor was asked to appear on BBC TV, along with Lyall Watson, in a program featuring the famous psychic Uri Geller. Taylor says his job was to be a "scientific hatchetman," but he was so amazed at Geller's performance that he launched into his own studies of paranormal phenomena. He described this work in his book *Super Minds* (Taylor 1973).

Taylor's companion on the Geller show, Lyall Watson, must also have been impressed by Geller's abilities, since Watson's credulous paean to the paranormal, *Supernature,* appeared shortly after the show (Watson 1973). Eventually, however, Professor Taylor had a change of heart, related in *Science and the Supernatural.* Here's what Taylor had to say:

> I started my investigations into ESP because I thought there could be something in it. There seemed too much evidence brought forward by too many reliable people for it all to disappear. Yet as my investigation proceeded that is exactly what happened. Every supernatural phenomenon I investigated crumbled to nothing before my gaze. . . . Error and deceit became more and more relevant for me in understanding the supernatural as my work proceeded. (Taylor 1980)

Professor Taylor could not bring himself to forgive his original critics, particularly magician James Randi, who had been telling him this all along. But he demonstrated other fine qualities in allowing the final judgment to be made by the facts. To my knowledge, Watson has never followed Taylor's lead in admitting his gullibility.

How to Demonstrate the Paranormal?

How can the promoters of paranormal phenomena prove they are both competent and honest? Simply, they must follow established scientific methodology. They will never succeed in gaining scientific dignity for their results if they insist on reformulating the scientific method to fit their preconceptions. Perhaps the scientific method will be reformulated someday; certainly it evolves with time. But to change both paradigm and method is asking a lot. The safest course for paranormalists is to follow the method as it is now codified. That is, they must: (1) convincingly rule out all conventional explanations, and (2) independently reproduce the reported effects. Following the scientific method is hard, agonizing work, and often fails. But it is not impossible.

Antimatter

Paranormalists often say that this standard is too difficult, that we conventional scientists demand more from them than we do from ourselves. This is simply not true. The method I have outlined is applied conventionally in science, and revolutionary new discoveries are still occasionally made. An important example of a revolutionary scientific discovery by the conventional method—and one that comes very close to what paranormalists seek to demonstrate—is the the mirror universe of antimatter. The discovery of this phenomenon opened up a whole new world that was previously not known to exist.

In 1928 Paul A. M. Dirac predicted that another type of matter exists that is a sort of mirror image of the matter of everyday experience. Now here was a phenomenon that certainly violated previous ideas at the most fundamental level. Yet scientists did not dismiss it for being too far out of the mainstream. They did not accept it, either, until being shown evidence.

Physicists looked for antimatter. Some claimed to see antielectrons, or positrons, produced by cosmic rays in detectors called cloud chambers. The tracks they observed had other, conventional explanations, though in most cases these explanations were rather unlikely. In 1932 Carl Anderson and Seth Neddermeyer were able to rule out other explanations.

In retrospect, we can see that many other investigators observed antimatter earlier than Anderson and Neddermeyer. But the latter are given credit for the primary discovery. Although they were not the first to see antimatter, their experiment was the first to show the positron hypothesis was the most likely explanation of the data—the least miraculous.

William James's White Crow

Perhaps America's greatest philosopher and psychologist was William James (1842–1910). He had an abiding interest in paranormal studies and helped found the American Society for Psychical Research (ASPR) in 1885. James was well aware of the many examples of fraudulent spiritualists and mediums, but argued that you only needed to find one who was not a fraud to prove the case for the existence of the paranormal. He wrote: "If you wish to upset the law that all crows are black, you must not seek to show that no crows are; it is enough to prove one single crow to be white" (James 1897).

William James was also the founder of the philosophy of pragmatism: "The true . . . is only the expedient in the way of our thinking, just as 'the right' is only the expedient in the way of our behaving" (James 1907). As others have paraphrased it: truth is what works.

So William James demonstrated the way to confirm the existence of the paranormal. Paranormalists simply have to find a white crow that works. After thousands of years of traditional beliefs, and over a hundred years of systematic study, we still await the white crow of the supernatural. As we will see, all the white crows put forward so far turn out to be black after all, or bogus, when examined up close. No doubt other candidate crows will be paraded before us in the future. I cannot imagine science ever being presented with an unequivocally testable proposition that will once and for all convince people to stop believing in a world beyond the senses.

References

Broad, W. J. 1979. *Science* 206:534.

Fairley, John, and Welfare, Simon. 1984. *Arthur C. Clarke's World of Strange Powers.* New York: G. P. Putnam Sons.

Feyerabend, Paul. 1975. *Against Method: Outline of an Anarchistic Theory of Knowledge.* London: New Left Books.

————. 1978. *Science in a Free Society.* London: Verso.

Fleischmann, Martin, and Pons, B. Stanley. 1989. *Journal of Electroanalytical Chemistry* 251:301.

Gardner, Martin. 1957. *Fads and Fallacies in the Name of Science.* New York: Dover.

Hines, Terence. 1988. *Pseudoscience and the Paranormal: A Critical Examination of the Evidence.* Buffalo, N.Y.: Prometheus Books.

Hume, David. 1748. *An Inquiry Concerning Human Understanding.* New York: Bobbs-Merrill. 1955.

James, William. 1897. *The Will to Believe and Other Essays in Popular Philosophy.* New York: Longmans, Green and Co. 1923.

————. 1907. *Pragmatism.* Cambridge, Mass.: Harvard University Press. 1975.

Kuhn, Thomas S. 1970. *The Structure of Scientific Revolutions.* Chicago: University of Chicago Press.

Langmuir, Irving. October 1989. "Pathological Science." Transcribed and edited by Robert N. Hall. *Physics Today.* P. 36.

Laudan, Larry. 1984. *Science and Values.* Berkeley, Calif.: University of California Press.

Popper, Sir Karl R. 1959. *The Logic of Scientific Discovery.* London: Hutchinson.

————. 1987. "Science: Conjectures and Refutations," in *Scientific Knowledge,* ed. Kourany, Janet A. Belmont, Calif.: Wadsworth. P. 139.

Randi, James. 1985. "The Columbus Poltergeist Case." *Skeptical Inquirer* 9-3:221.

Rawcliffe, D. H. 1959. *Illusions and Delusions of the Supernatural and the Occult.* New York: Dover.

Rothman, Milton A. 1988. *A Physicist's Guide to Skepticism.* Buffalo, N.Y.: Prometheus Books.

Taylor, John. 1975. *Super Minds.* New York: Viking.

————. 1980. *Science and the Supernatural.* New York: Dutton.

Theocharis, T., and Psimopoulos, M. 1987. "Where science has gone wrong." *Nature* 329:595.

Watson, Lyall. 1973. *Supernature.* London: Hodder & Stoughton.

4.

From Magic to Science

Man became free when he recognized that he was subject to law.

Will Durant

People without Religion

Early in the first volume of his monumental work, *The Story of Civilization,* Will Durant notes that some people have no religion. For example, Vedah tribesmen in Ceylon, when asked about God, answer: "Is he on a rock? On a white-ant hill, on a tree? I never saw a god!" A Zulu is asked: "When you see the sun rising and setting, and the trees growing, do you know who made them and governs them?" The answer: "No, we see them but cannot tell how they came; we suppose that they came by themselves" (Durant 1935, p. 56).

These tribesmen instinctively know what Western science, after centuries, has still failed to impress upon the average "civilized" person: the evidence of your two eyes and other senses far outweighs that of any external authority, no matter how exalted.

Fear of Death

So why do most people fervently believe what their eyes fail to confirm? Durant attributes it to fear, particularly the fear of death. He agrees with the Roman poet Lucretius that fear is the mother of gods. Even today, the promise of eternal life remains the primary appeal of religion. Intellectually we are forced to accept the ultimate destruction of our physical bodies, but we still look for a way out of the inevitable. Accepting that our material bodies will not survive forever, we hold out hope that some immaterial part of us—soul, spirit, psyche, mind, consciousness—is eternal.

Dreams

We can never know for sure what went on inside the heads of our prehistoric ancestors. Durant speculates on how awareness of death gradually developed, and with it concepts of the supernatural that evolved into the familiar forms of today. Death would have been very puzzling and disturbing to the primal mind (as it is indeed to the modern mind). The suddenness and permanence of death must have seemed strange and unnatural to the first people to develop the intelligence to think about it. Durant notes that "Primitive life was beset with a thousand dangers, and seldom ended in natural decay; long before old age could come, violence or some strange disease carried off the great majority of men. Hence early man did not believe that death was ever natural; he attributed it to the operation of supernatural agencies" (Durant 1935, p. 57).

Besides the puzzle of death, early people must have been frightened and threatened by mental pictures. Awake and in dreams, the most fantastic visions appear in our heads. Durant speculates that "Primitive man marveled at the phantoms that he saw in his sleep, and was struck with terror when he beheld, in his dreams, the figures of those whom he knew to be dead" (Durant 1935, p. 57). A dead person could still be pictured, his voice heard. He was a spirit. But was he dead or not?

A dead body disintegrated, while the thought and dream images—the spirit—of the individual who had occupied the body lived on. So the dead began to take on special meaning, eventually to be worshipped. As Durant notes, the word for god among many primitive peoples actually means "dead man."

The Origin of Gods

We take for granted that we are more sophisticated thinkers than our ancient predecessors. Our languages, from East and West, contain many thousand words used to classify the countless phenomena of our experience. This enables us to go beyond the obvious distinctions of earth and sky, grass and trees, humans and animals. We make more abstract divisions of phenomena into real and imaginary, body and spirit, living and nonliving, natural and supernatural.

If our ancestors were anything like the modern people in remote isolated tribes that we call primitive, they would not have organized phenomena into the same or even the same number of cubbyholes as we do. The imagined, the dreamed, the hallucinated would have been as real as the perceptions of eye and ear we call real. The natural and supernatural, body and spirit, would have been indistinguishable.

If our ancestors thought like the Vedah tribesmen quoted earlier, they would tend to believe their eyes—those inside their heads as well as those outside. Humans and animals would be seen to move amd have life. But so would the water in a stream, a cloud in the sky, or a rock tumbling down the side of a hill. They would all be alive.

Causal relationships would be inferred, especially in regard to movement. The notion of force as the instrument of motion would be dimly perceived, and if that force were not seen, then it could be imagined. Further, the unseen force would not be something beyond the realm of normal experience, like the gluon exchange physicists now use to describe the interaction between quarks; rather the cause of motion would be personal, living, human.

So invisible living entities with human characteristics, gods of earth and sky, were invented to account for movements. The weather, changes in seasons, growth and decay, were caused by these gods. Gods or spirits were inside rocks, trees, rivers, wind, and animals, providing the changing, rhythmic motions of a living world. And these gods were inside people's heads too, directing them on what to do.

Fermilab in the Sky

We see in these perceptions the first appearance of a kind of science— the seeking of causal explanations for the phenomena of the world. In today's physics classes, the first thing students learn about is motion. However, they are taught that motion at constant velocity does not require the action of a force; a force is needed only for changes in velocity, acceleration.

This observation is expressed by Newton's second law of motion: $F = ma$, where F is the force on a body, m is its inertial mass, and a is the resulting acceleration (the rate of change of velocity). In other words, movement does not require a mover, only a change in movement does. And if no mover is seen, then one does not have to invent one that is unseen. The ultimate prime mover, which Aristotle and Aquinas defined as God, is not required by the data. But perhaps we should allow God to be redefined as the prime accelerator—Fermilab in the sky.

The Beginning of Science

So the primitive mind started on the road to science when it began developing causal concepts. In *The Golden Bough,* Sir James Frazer argues that science had its roots in the magic of primitive animism. Early shamans and priests were basically magicians whose job it was to interpret and demonstrate the powers of the invisible spirits. The rituals were not always successful, so the magician helped the supernatural along with a degree of knowledge of what we now call natural forces. Of course this knowledge was kept secret; its effects were attributed to the supernatural. But out of it grew astronomy, chemistry, medicine, and physics.

Science is not, as most people think, a recent development of the human mind, springing up in the sixteenth century out of a vacuum. Rather it is as old as conscious thought itself, with roots in charlatanism and pseudo-science. As Will Durant said, "Magic begins in superstition and ends in science" (Durant 1935).

Myth and Science as Metaphors

Many writers regard ancient myths as allegories or metaphors for deeply perceived concepts of reality. However, the myths of the ancients were as real to them as the rocks and trees around them. Only with the development of abstract thought did the explanations for phenomena move out of the realm of everyday experience to invisible causes of events.

Science deals with many notions that are not common sense, that go beyond everyday experience. Is it more sensible that invisible photons interacting with atoms and electrons result in a lightning bolt, or that some god tossed the bolt across the sky? What would a Vedah tribesman say about quarks and leptons? I could take him to the lab and show him the trails of bubbles in a bubble chamber photograph that convinces me of the existence of the electrons and other particles that left the trails, but would he be convinced? I greatly doubt it, unless I could relate the electrons and photons to his own experience.

Today's quarks and leptons can be viewed as metaphors of the underlying reality of nature, though metaphors that are objectively and rationally defined and are components of theories that have great predictive power. And that's the difference between the metaphors of science and those of myth: scientific metaphors work.

We accept scientific explanations today, not because they are in any way more sensible than superstition and myth, but because these explanations provide far more ability to control and manipulate nature. Despite the insensibility of the mathematical equations and abstract models of science, they provide us with a superior description of reality. Unlike myth, we can put the equations to work in predicting future events, and thereby exert some control over nature. We may be uncomfortable with the fact that science has shown us how to destroy a city with one bomb. But the fact is, no supernatural metaphor has any such power, except in the movies.

In the pragmatic view of truth of William James, science is true because it works. Science may not be the only path to the truth, but it is the best one we have yet been able to discover.

From Magic to Myth

In *Masks of the Universe,* Edward Harrison makes a distinction between magic and myth. In the magic universe of the first thinking humans, the activating agents of the world are creations of the mind but still a part of that world. This is, in a way, quite like modern science. Only now we have particles and symmetries in place of spirits. At some point in the past, the inner mental entities of our ancestors were separated from their original physical forms. They become superphysical, metaphysical (Harrison 1985).

Animistic spirits ultimately evolved into the cosmic gods of the Nile and the Tigris-Euphrates. The pulsating, living world of the magic universe disappeared and was replaced by the mythical universe peopled by spirits that no longer resided in nature. Thus began religion as we know it today, in which ultimate reality transcends the visible, physical world. Ironically, this backward step happened with the advance of civilization: with the invention of agriculture and then that most terrible of all human inventions— the city.

Humanity Becomes Social

As people clustered together, first in villages and then in cities, society had to become organized. Instead of roaming freely, either individually or in small bands, in search of game, men had to stay in one place to work the fields and to barter for their other needs; they and their mates were no longer able to supply all of their needs for themselves. Soon the new economy of trade developed, which forced people into greater dependencies on one another. Humanity became social. Leaders were now needed to keep some kind of civil order, and the village shaman and temple priest, with their supposed supernatural powers, proved to be effective in keeping everyone in line.

We can imagine how the gods of the shamans gradually grew more remote from the people, making the shaman more awesome and enhancing his power. Based on the real power of the sword, kings and empires came into being, and the glory of kings provided a new cosmic model for the imagined gods.

The transcendent reality took on different forms in Orient and Occident. As Joseph Campbell points out in *The Masks of God: Occidental Mythology,* the "ultimate ground of being" in the Orient transcends thought itself and is beyond rational consideration. In the Occident, this ground of being is personified as a Creator. Two kinds of pieties, Eastern and Western, develop: the *religious* pieties of Zoroastrianism, Judiasm, Christianity, and Islam; and the *humanistic* pieties of the Greek, Roman, Celtic, and Germanic (Campbell 1964).

The Great Goddess

When human villages were small agricultural communities, the primary deity was the earth itself. Since the earth brought life, this deity was personified as female. With the rise of the first great city at Sumer (c. 3500–2350 B.C.E.), the earth goddess of the neolithic village evolved into the Great Goddess that Campbell calls "a metaphysical symbol: the arch personification of the power of Space, Time, and Matter." In the great civilization of Egypt, Isis was the mother of all things. And for the earliest Hebrews, probably Eve was mother goddess and Adam her son. The more familiar tradition of Eve's creation from Adam's rib is, according to Campbell, an inversion resulting from the later male dominance that occurred with the rise of the warrior kings and their cruel empires of the sword (Campbell 1964).

God as King

The invincible sword was the product of the iron age (c. 1250 B.C.E.) and with it came the dominance of the warrior. It would not do, of course, to have a female deity ruling over a majestic king, so the Great Goddess was replaced by supreme male gods. In Greek myth, Zeus defeats Typhon, child of Gaia (goddess Earth), securing the dominance of the patriarchal gods on Mount Olympus over the Titans, the children of Gaia. Babylonian myth provides a parallel in Marduk's defeat of Tiamat.

But even in these cases, the female originally gave birth to the world. In the Hebrew Bible, however, the transformation to male dominance becomes complete—a male god creates the universe unaided by any consort, although the Christians would later provide him one in the Virgin Mary.

And so we can trace the development of the idea of the Father/King/

Creator of the Judaeo-Christian-Islamic myth. The animistic spirits of the hunter evolve into the earth deities of the farmer, and then into the Great Warrior King who rules heaven just as earthly domains are ruled by their kings. He is an invisible, invincible, frightening god, but not unknowable. God speaks to certain men, who then relay his orders to the rest of humankind. With the development of writing, his word is transcribed into scriptures that then become the source of authorized knowledge of the supernatural realm.

Beyond Comprehension

In the Orient, a somewhat different view of ultimate reality develops. A supreme reality exists beyond the reality of the senses. But it is immensely beyond comprehension, unknowable by the finite human mind. Here for the first time we can rely on written records, rather than inference, for insight into the thinking of ancient people. Other than a few fragments from Egypt, the *Upanishads* of India are the oldest existing written philosophical tracts of the human race. There we find the message that "the material world (is) nothing but a long-enduring kaleidoscopic deception, the only thing that is worthy of man's aspirations is transcendence" (Sinari 1970).

Further, the human intellect is incapable of understanding something so immense and so complex as the transcendent reality. Thus, it is not by learning, thinking, reasoning that one gains insight into trancendent reality—states called *Atman* and *Brahmin*—but by cleansing the mind of all thoughts and worldly aspirations.

Unlike the more hopeful philosophies of the West, the teachings of the ancient Indian sages led Indians to find "the world to be continually weary, an unwanted hindrance, a spoke in the wheel of their trans-phenomenal ascent, a bondage" (Sinari 1970). In other words, a philosophy in which the material world is of negligible importance is not of much help in getting by in the material world.

The Natural Order

Fortunately, the Indian view of the futility of intellect and body did not penetrate to the Aegean coast of what is now Turkey, then dotted with colonies of Greeks who had fled their homelands ahead of invaders from the north. There, for a brief time, a few individuals were rich and secure enough to contemplate something other than how to get their next meal or avoid massacre by their neighbors.

Astronomers calculate that on May 28, 585 B.C.E., a total eclipse of the sun occurred that would have been visible from this region. Perhaps this eclipse was the one that Thales of Miletus is said to have predicted from Babylonian tables. In any case, about that time Thales became the first human on record to speculate that a natural cosmic order existed that allowed prediction of natural occurrences such as eclipses.

To Thales, water was the basic stuff of the universe, the cause and guiding principle behind all existence. Everything came from water. The most important feature of Thales' concept was that here, for the first time, the basic principle was not some invisible spiritual force, but everyday stuff that we can all see, touch, and taste. It was real, natural.

The new idea was that behind sensible phenomena there exists no personified spirit acting on its own capricious will. Rather, the phenomena happen in a predictable way according to natural law, which still distinguishes the scientific from the religious approach to explaining the universe. If applied science began with the shamans, theoretical science began with Thales and his prediction of the May 28, 585 B.C.E., eclipse.

Most historians flag Thales as the first philosopher of the West. As Durant says, "Here for the first time thought became secular, and sought rational and consistent answers to the problems of the world and man. . . . Here was the idea of law, as superior to incalculable personal decree, which would mark the essential difference between science and mythology, as well as between despotism and democracy" (Durant 1939, p. 135). The belief in supernatural forces remains to this day a yoke on the neck of humanity, but at least Thales made it possible, for those of us who wish it, to be free of that yoke.

The Reductionists

Thales' concept of a natural explanation for the universe was not immediately adopted beyond his school in Miletus. But then, a century and a half later, Empedocles of Sicily (c. 495 B.C.E.) made a number of concrete improvements on the vague ideas of Thales that begin to sound like modern science. First, he said that all objects are composed of a small number of basic elements: earth, air, fire, and water. Not quite ready to discard the concept of soul, he suggested that soul was also made of these elements.

Empedocles recognized that matter passes through solid, liquid, and gaseous stages. Further, he proposed how the elements interacted: by the forces of Love and Hate, which we can recognize as a poetic way of saying attraction and repulsion. And important in terms of our ultimate understanding of normal sensory perception, he suggested that emanations thrown off from bodies penetrated the senses, providing the mechanism for sensation.

Of course, the detailed processes that Empedocles described were wrong, but his basic approach to understanding nature—reducing it to the simplest elementary objects and forces—is largely the one followed today by much of science. This reductionism is a key ingredient in our modern models of the universe.

Today particle physicists see everything as quarks and leptons interacting by the exchange of photons, gluons, and other gauge bosons. Most people find it difficult to accept this view as the complete story. They intuitively feel that the whole can be greater than the sum of its parts, as a cathedral is more than a pile of rocks or a person is more than a collection of atoms.

Collective phenomena such as heat or condensation exist, of course, but these are fully understood from a consideration of the statistics of the large numbers of particles involved. However, the more complex collections of particles that make up the systems of our normal experience, from polymers to people to planets, have properties that emerge as the result of their organization. Are these so-called emergent properties inconsistent with the reductionist view? Do holistic forces exist that can cause violations of the laws of microscopic physics? A later chapter will be fully devoted to this issue, which is fundamental to the theme of this book.

The Atomists

The elements of Empedocles, like those of nineteenth-century chemists, were not invisible particles but the perfectly observable materials earth, fire, water, and air. The concept of atoms—solid indestructible fundamental objects too tiny to see, moving around in an otherwise empty void—was first formulated by Leucippus (c. 435 B.C.E.). However, little is known of him and credit is usually given to his associate Democritus (c. 470 B.C.E.), from whom some important writings have survived. In his writings Democritus teaches that all things, even the soul, are atoms. No supernatural agency exists. He even goes so far as to argue that belief in divine power resulted from the apparitions seen in sleep and alarming aspects of nature like lightning, thunder, earthquakes, and floods.

The Materialists

Alfred William Benn has nicely described the contributions of the materialists, who sound so modern to our ears: "They first taught men to distinguish between the realities of nature and the illusions of sense; they discovered or divined the indestructibility of matter and its atomic constitution; they taught that space is infinite. . . . They held that the seemingly eternal universe was brought into its present form by the operation of mechanical forces. . . . They declared that all things had arisen by differentiation from a homogeneous attenuated vapour." Their central doctrine was "that the universe is a cosmos, an ordered whole governed by number and law, not a blind conflict of semi-conscious agents." If not science yet, these ideas made science possible, presenting a theory of the universe that "methodised observation has tended to confirm" (Benn 1982, p. 5).

At about the same time in India, a school of Hindu materialists called the "Charvakas" were saying remarkably similar things. They saw no evidence in the world for Atman, for demons, gods, or soul. The one reality was matter, atoms. Mind was matter. They argued that notions of immortality and rebirth were fraudulent, that morality was social convention (Durant 1935, p. 418).

The Founders

Unfortunately, these eminently rational men of East and West did not win over the minds of those thinkers in either region who turned out to be most influential in the history that followed. Living at about the same time as Thales, give or take a generation, were the founders of some of the world's great religions: Zoroaster in Persia, Lao-Tze and Confucius in China, and Buddha in India. Equally influential were the Hebrew scholars then in exile in Babylonia, who put their tribal myths on paper and in the process incorporated the creation and flood myths and other legends of their hosts.

The Pythagoreans

Also at about this time, in southern Italy, Pythagoras (c. 569–c. 507 B.C.E.) founded a brotherhood that provided yet more impetus toward the ultimate victory of supernatural belief in the ancient world. The Pythagoreans developed the view that all things are number, or made out of numbers.

Now this sounds scientific doesn't it? Indeed, where would science be without the Pythagorean Theorem and the other mathematical developments that came from the Pythagorean school? But the Pythagoreans did not simply say that nature is described by numbers, as we do in modern science. Rather, to the Pythagoreans number is the absolute principle of existence: a spiritual power that cannot be seen, like fire or water, but that controls the sensory world.

So a new mystical concept was introduced that was far more sophisticated than the other superstitions of the time. Unlike the materialists, who viewed the soul as being the same stuff as objects in the sensory world, or the animists who made no distinction between spirit and matter, the Pythagoreans introduced into the thinking processes of humanity the separation of soul and body. This duality haunts us to the present day. Further, the notion was introduced that the world of our senses was not the world that really mattered; beyond the sense world was the true, perfect reality to which we should aspire.

The appeal of a world beyond the senses was enormous, and remains so today. The offer of escape from the suffering and imperfection of life, the promise of our participation in some grand cosmic consciousness, are almost impossible to resist. But what if dualism is wrong, just an enduring fantasy of the childhood of humankind? Then what good does it do for us to live that fantasy, to deny reality?

The Dualists

The doctrine that only one ultimate substance or principle exists is called "monism." The idea that everything is matter and nothing more is "materialistic monism." Thales' proposition that water is the source of all existence was the earliest form of materialistic monism. Directly opposed to materialistic monism is the view that everything that exists is an idea of the mind; this form of monism is called "idealism."

The view that two forms of reality exist is called "dualism." Dualism began in the West with the Pythagoreans, was later adopted by Parmenides, and culminated in Plato's theory of forms. But dualism had been formulated much earlier in the East, in the Hindu *Upanishads*.

Perhaps the idea of dualism trickled into the Hellenic world from the East. Some communication between East and West surely existed in ancient times. A generation after Plato, Alexander the Great would conquer his way to India. So Mediterranean people obviously knew of that ancient land. According to Will Durant, "Pythagoras, Parmenides and Plato seem to have been influenced by Indian metaphysics; but the speculations of Thales, Anaximander, Anaximenes, Heraclitus, Anaxagoras and Empedocles not only antedate the secular philosophy of the Hindus, but bear a sceptical and physical stamp suggesting any other origin but India" (Durant 1935, p. 533).

The Athenians

Most of what we know about the early Greek thinkers has come down to us from those philosophers whose teachings have survived in the greatest quantity: Plato (c. 427–347 B.C.E.) and Aristotle (384–322 B.C.E.). The sheer bulk of these writings, if nothing more, resulted in the ideas of Plato and Aristotle having a dominant influence upon future generations of philosophers. From Plato we also learned about the first member of the illustrious Athenian philosophical trinity, Socrates (469–399 B.C.E.).

Socrates' great contribution was to teach us the value of clear thinking and skepticism. His barbs were devastating to the "open-mindedness" of the fashionable sophists who applied democracy to the world of ideas in a way not much different from some of today's critics of science.

Socrates was the kind of down-to-earth commonsense kind of guy that we admire so much in America. He disdained the cosmic issues that concerned other philosophers. Feeling that philosophical speculation had gone too far, he said men should focus on human affairs and forget theology, physics, and metaphysics. I find that a shame. I would have loved to have been able to read what Socrates' great mind would have made of the ultimate questions. Or, perhaps we already know: he may have decided they were a waste of time.

His student Plato eloquently passed on to us Socrates' words and ideas, but did not heed the Master's call to ignore metaphysics. Turning back to the idealism of Pythagoras and Parmenides, and rejecting the empiricist notion that the senses are the only source of knowledge, he said that the true reality was "Ideas" or "Forms."

These objects of thought were immortal, living beyond the life of the one who thought them. For example, a geometrical figure drawn in the dirt with a stick will be crooked and imperfect, and can be wiped into oblivion with a sweep of the foot. But the idea of the straight line, circle, or triangle was perfect and everlasting. The laws of geometry applied to the ideal geometric figures, not the ones sketched in the dirt.

The Scientist

Plato's student and the mentor of Alexander the Great, Aristotle, did us an enormous service by returning to the senses as the primary source of knowledge, and setting the pattern for science by establishing systematic observation and experimentation as the foundation of scientific method.

In biology, Aristotle usefully consolidated much previous work, but rejected Empedocles' prescient intuition of what we now recognize as the most important principle of biology: evolution by natural selection. Likewise, in physics, Aristotle failed to grasp the primary principles of motion. While he helped clarify many physcial concepts, Aristotle's idea of absolute space and the need of a force for motion were totally wrong.

By the Middle Ages, Aristotle's authority had become so great that progress in science was severely retarded by an adherence to Aristotelean concepts that simple experimentation could demonstrate as grossly incorrect. For example, Aristotle taught that objects fall at speeds in proportion to their weights, a false notion that can easily be shown to be incorrect by the simplest experiment. Only when Galileo and his contemporaries finally asserted the authority of observation over that of revered ancient texts did progress in science finally pick up where it had essentially left off after the time of Aristotle.

Equally significant to the historical development of human thinking, Aristotle formulated metaphysical ideas that were later to be used by Church theologians, especially Augustine and Aquinas, to provide Christianity with a systematic theological structure. In his metaphysical ideas, Aristotle follows Plato's lead. He makes invisible forms or ideas, rather than detectable matter, the essential component of being. Aristotle defined God as the form of the world, the First Cause—Uncaused, the Prime Mover—Unmoved. Thomas Aquinas (1225–1274) would use this same definition in his "proof" of the existence of God.

But, as we have seen, movement does not require a mover, and modern quantum mechanics has shown that not all effects require a cause. And even if they did, why would the Prime Mover need to be a supernatural anthropomorphic deity such as the Judaeo-Christian God? Why could it not just as well be the material universe itself?

Naturalism and Supernaturalism

So several ancient Greek philosophers provided a viable alternative to belief in a supernatural world. But supernatural beliefs were buttressed by other philosophers who became more influential in later history.

The conflict between the two points of view could not be clearer. In the natural view, our only source of knowledge about the world outside our heads is the senses. This includes any knowledge that may be programmed in our genes, since such knowledge would have originated with the sensory inputs received by our ancestors during a billion years of evolution.

It can be argued that other forms of knowledge exist, such as pure mathematics and symbolic logic, that seem to have no dependence on sensory data. If so, then that knowledge resides within the collective consciousnesses of mathematicians and logicians, and need not relate to anything beyond those realms, although this remains a debated point. As they relate to the external world, reason, logic, and mathematics are thinking tools we have developed to describe the observations of our senses. They are not a part of the real world, but are inventions of the human mind, like language. Still, if these tools are inventions, few would deny what these tools describe: objects in the world, which are most immediately known through our senses, and which have some quality we commonly regard as real.

In the view that we can call "supernatural," ultimate reality is beyond the senses. We learn about it through nonsensory, mystical mechanisms: revelation, reason, meditation, pure thought. This is clearly a dualist view, featuring a mental or spiritual component that is beyond body, and special channels to or even direct participation in a cosmic consciousness. Despite thousands of years of progress in human thinking, and the daily demonstration in modern society of the power of materialistic science, the dual universe of matter and spirit remains the unquestioned belief of the great majority of the human race.

Why Did Life Form? Why Not?

How did the human race come to be, if not by the creative act of a transcendent power or being beyond our comprehension? If no such being exists, then why did life form? My answer is another question: Why not?

It often happens in science that the absence of a fact has to be explained, not the existence of one. For example, the great principles of physics, such as the first and second laws of thermodynamics, forbid certain events from taking place. However, they mandate none.

And so with life. When the conditions were right on earth a few billion years ago, atoms floating about in the air or sea stuck together in great numbers to produce large, complex molecules. Just by chance, some molecules happened to develop the ability to copy themselves, using energy and material in their vicinity. So, where originally one molecule existed, soon there were two, and then 4, 8, 16, 32, 64, 128, 256, 512, and so on, in geometric progression.

Only the limitation of available material and energy prevented the world, in a very short time, from being thickly coated with exact duplicates of that original reproducible molecule.

For example, suppose the primordial molecule was a sphere one micron in diameter (10^{-6} meter, less than a ten-thousandth of an inch) that reproduced at the rate of once per second. With unlimited resources, in only 145 seconds the earth would have been covered with a layer of molecules out to the orbit of the moon, 240,000 miles away. If you think that's fantastic, in another 174 seconds of unlimited reproduction, the micron-sized molecules would fill the universe. Such is the power of the geometric progression. In 319 generations, each of which doubles the number of molecules, the original molecule would produce 2^{319} or 10^{96} new molecules.

But this molecular population explosion was stopped, for the same reason the current human one will eventually be stopped, by limited resources. The molecules quickly leveled off to a large but finite population. Not much else would have happened except for another fact: each molecular copy was not always perfect. An occasional ultraviolet photon from the sun, cosmic ray muon, or other stray energetic particle would knock a random atom out of place in the structure of an individual molecule, or cause the shape of a molecule to skew one way or another. Alternatively, a copying error might occur. These were all chance events, but nevertheless served to produce a variety of structures in the place of the original one, as the new feature appeared in succeeding copies.

If the new feature made it more difficult for the new species of molecule to be copied further, that species became extinct. Most changes were probably neutral. But at least some of the new molecular structures were more successful at copying themselves than others. These species had a distinct advantage, and quickly tended to dominate over less efficient structures in their mutual struggle for the same limited resources.

And so, Darwinian evolution by natural selection began. Undoubtedly, the original structures were simpler and less efficient than the present DNA-based genetic patterns. But eventually the present genetic scheme developed, along with sex for the larger structures, simply because these features facilitated reproduction and enhanced survival. And so, living organisms evolved. The fundamental driving force within them was simply the preservation of the information needed to reproduce faithful copies of themselves.

In each of our cells, information exists that enables the structures of our complex bodies to be maintained, as the atoms that compose our cells are continually replaced. The particular particles of matter that make up our bodies at a given time do not define us as individual organisms. Instead, our true self is the information coded in our DNA.

As Richard Dawkins says in his brilliant and highly entertaining exposition of the natural processes of evolution, *The Blind Watchmaker,* "What lies at the heart of every living thing is not a fire, not warm breath, not a 'spark of life'. It is information. . . . If you want to understand life, don't think about vibrant, throbbing gels and oozes, think about information technology" (Dawkins 1987).

The Evolution of Religion by Natural Selection

We have attempted to understand how the idea of a world beyond the senses might reasonably have entered human thinking. However, our understanding of the basis of the supernatural is still incomplete. If supernatural beliefs were simply the product of the unsophisticated thinking patterns of early humans, then they should have largely faded away in our scientific age. Yet every survey of people's beliefs continues to indicate a strong majority who believe in God, angels, the devil, astrology, and various other occult and supernatural phenomena. As philosopher Paul Kurtz has said, "The transcendental temptation lurks deep within the human breast. It is ever-present, tempting humans by the lure of transcendental realities, subverting the power of their critical intelligence, enabling them to accept unproven and unfounded myth systems. . . . Any impartial observer of human history must be duly impressed by the tenacious character of religious faiths and the persistence of transcendental myths in human history, as well as the ability of the most outrageous myths to sprout and take root, even in the graveyards of defeated and dying systems. Is this a perverse strain, like original sin, but beyond any hope of human redemption?" (Kurtz 1986).

Any attempt at understanding humanity must include an explanation of the hold that supernatural belief continues to have on most of the human race. Here again we can call upon materialism for a reasonable hypothesis: religion evolved by a process analogous to the natural selection that produced us and every other living species. Religious belief may now be deeply programmed in our DNA. This may have happened because, at one time, such beliefs provided a survival advantage for the people who had such coded information in their genes.

In the early days of the human race, we were few in number and struggled in competition with other species to survive. In that precarious situation, special advantage would accrue to those people living in communities with strong tribal rules forbidding behavior patterns threatening to the survival of the community. These ranged from taboos against incest and murder to special dietary prohibitions.

By attributing these taboos to supernatural command, the leaders and their priests could enforce them more effectively. Individuals with a genetic disposition to question or disobey the rules would be suppressed, ostracized, or even killed. So they were less likely to pass their skeptical genes on to the next generation. As William Simms Bainbridge has put it: "The human mind did not evolve in order to create a race of philosophers or scientists" (Bainbridge 1988).

Now the way evolution works—by means of random events—does not require the process to be perfect. This is an often-misunderstood feature of evolution. Opponents of evolution argue that if you can find a single counter example, you destroy the whole edifice. Not at all. A genetic characteristic leading to a slightly higher probability for survival than others will generally lead to that characteristic becoming more common as the generations progress. So, the set of particular mental characteristics that led to greater credulity, a slightly greater willingness to believe the preposterous in the face of all evidence to the contrary, was sufficiently enhanced to remain within our genes to the present day. But it need not be so strong in each of us that it cannot be overcome.

Overcoming Our Instincts

The genetic programming in favor of supernatural belief does not have the value for survival it once had—in fact, quite the contrary. Where once human tribes needed as many members as they could produce in order to survive, the global human tribe today finds its own survival threatened by too many members. And most of these members still hold to the outmoded beliefs that may have helped their ancestors, but hinder us all today. These beliefs threaten our survival rather than aid it, as they mitigate against birth control and the other measures needed to halt the exponential increase in our population.

Believing, despite the overwhelming evidence from biology and astronomy, that humans were made in the image of God and given dominion over the earth, we in the Western World have also found it easy to pollute the earth and wantonly destroy many of its life forms. In the Judaeo-Christian ethic, human life is taken to be sacred. But not all life. Only humans have souls.

Evolution does not guarantee the survival of a species. In fact, it dooms most species. Most of the species that have existed on this planet are now extinct. And our natural instincts, if left to themselves, will probably result in the extinction of our own species before long.

Fortunately, by the same process of evolution, we humans have developed a unique quality that gives us the power to overcome our instincts. This power resides in our intellect. Only through the application of intellect to overcome the dangerous behaviors programmed in our genes can we expect to survive. And our intellect is pointing the way for us to reprogram our own genes, to rid them of the transcendental temptations that now threaten our very existence.

References

Bainbridge, William Simms. Spring 1988. "Is Belief in the Supernatural Inevitable?" *Free Inquiry* 8:21.

Benn, Alfred William. 1982. *The Greek Philosophers.* London: Kegan, Paul, and Trench.

Campbell, Joseph. 1964. *The Masks of God: Occidental Mythology.* New York: Viking.

Dawkins, Richard. 1987. *The Blind Watchmaker: Why the Evidence of Evolution Reveals a Universe Without Design.* New York: W. W. Norton.

Durant, Will. 1935. *The Story of Civilization I: Our Oriental Heritage.* New York: Simon and Schuster.

———. 1939. *The Story of Civilization II: The Life of Greece.* New York: Simon and Schuster.

Frazer, Sir James. 1962. *The New Golden Bough,* abridged, edited, and revised by Theodor H. Gaster. New York: Mentor.

Harrison, Edward. 1985. *Masks of the Universe.* New York: Macmillan.

Kurtz, Paul. 1986. *The Transcendental Temptation: A Critique of Religion and the Paranormal.* Buffalo, N.Y.: Prometheus Books.

Sinari, Ramakant A. 1970. *The Structure of Indian Thought.* Springfield, Ill.: Charles C. Thomas.

5.

Is There a Mystical Path
to a World Beyond?

*Men think epilepsy divine, merely because they do not understand it. But if
they called everything divine, why, there would be no end of divine things.*

Hippocrates

The Stigmatist

In 1224, Francis of Assisi ended his administration of the religious order
he had founded fifteen years before and secluded himself in a deep ravine
on Mount Verna. There, on September 14, after a long period of fasting
and prayer, he saw a vision of the crucified Christ. Immediately afterward
he felt pains in his hands, feet, and side. Looking, he saw what appeared
to be hemorrhages under the skin at the traditional locations of Christ's
wounds—the stigmata.

The image of Christ nailed to the cross is presented to us in countless
paintings and carvings. But these depictions are largely the products of
the imagination of the artists who created them, centuries after the event.
No one really knows the details of how Christ was actually crucified. Most
likely he was not nailed, but tied with rope to the cross, the usual method
used by the Romans of the day.

So how is it that Saint Francis of Assisi and many other stigmatists
received the stigmata in the traditional but probably not actual locations
of Christ's wounds? The answer lies in the close interaction between mind

and body or, more accurately, brain and the rest of the body. We control our bodies with our brains, consciously and unconsciously. We accept without question and attribute no special significance to the brain's control of everyday fuctions, such as walking, talking, and breathing. But with respect to unusual and abnormal functions—that is, when unusual and abnormal things happen—we have a tendency to assign some deeper meaning, looking for causes beyond the forces of everyday experience.

Mind over Matter

Author Norman Cousins has effectively promoted the notion that positive influences, such as laughter, can help cure serious illnesses. He believes that he cured himself from a rare disease by "love, hope, faith, laughter, and confidence" (Cousins 1976), although exactly how life-threatening his illness actually was is questionable (Butler 1985, p. 641).

Cousins has joined the staff of the UCLA School of Medicine, where studies indicate that states of mind may stimulate the production in the body of various healing chemicals. However, neither Cousins nor the researchers claim such beneficial effects are in any way supernatural or miraculous. And their work is certainly not conclusive; the results can still be interpreted in terms of the placebo effect.

If the normal and usual are taken to be natural, the unusual also can be natural. We do not seek paranormal explanations for snow in June or a supernova. Why should we need them for unusual mental happenings? Under emotional distress, we can give ourselves ulcers and headaches. Psychosomatic illness is clinically well-established, though probably not as common as even many medical doctors think. Doctors are not immune from the human tendency to attribute extraordinary explanations to events when often a simpler, more economical, but less exciting explanation will suffice.

Considerable evidence exists for a certain limited healing power of the brain. But, why not? The brain is wired to the rest of the body, and if it can control our motor functions, why should its control over other functions crucial to our well-being be so hard to accept? Surely evolution would have provided appropriate mechanisms. Nothing about the brain-body interface requires it to be in any way spiritual, beyond explanation by natural means. However, the healing power of the mind is often greatly exaggerated, simply because everyone—doctors and patients—wants to believe in it.

Psychiatrists who have examined stigmatists find characteristics of hysterics. And despite so many stories of stigmata, official Catholic sources have no list of living stigmatists. Clearly our mental condition can affect the flow of blood to various parts of the body and cause other physical changes. Our complexion turns red when we are embarrassed. When we are afraid, we turn white and start to sweat, our mouth becomes dry, and our blood pressure increases.

Most of these effects can be brought under control by biofeedback. Psychiatrists have observed bleeding and blister formation in patients and attribute it to a combination of hysterical predisposition, heightened suggestibility, and hypnotization. In the past, researchers have even been able to trigger such phenomena with certain stimuli, but today this type of experimentation with humans is considered unethical (Zusne and Jones 1982).

The Faith Healers

In recent years we have seen a rise in the attempted use of nonmedical forms of healing, as people refuse to accept the admitted limitations of science. They are disappointed that science can put people on the moon, but cannot make them live forever. TV evangelists, including one 1988 presidential candidate, have claimed healing powers that they attribute to supernatural forces.

Magician James Randi has shown that many faith healers are actually charlatans who cynically prey on the hopes of the ill and crippled. They gain riches for themselves, but the consequences are often disastrous for their victims, who may forgo conventional treatment (Randi 1988). Many of these unfortunates die terrible deaths, with the added anguish of blaming themselves for their lack of faith, as demonstrated by the failure of the faith-healing program.

Of course belief in faith healing is not new. Cures have traditionally been attributed to holy relics or to touching the hem of a garment worn by a king or queen. Holy men and women in most religious traditions are said to have effected miracle cures. I don't mean to imply that all historical religious healers were conscious frauds or that some current practitioners are not sincere; I simply question the evidence that such powers exist. On close examination, this evidence is flimsy, indeed.

Each year, many thousands of people visit Lourdes hoping for miraculous cures. The city has been converted into a huge commercial trap, what TV's "60 Minutes" called "curios rather than cures." The city fathers claim 30,000 healings per year, though the Catholic Church has certified only sixty-four miracles over all the years since Lourdes became a shrine. Of course, the Church believes in miracles; none of these sixty-four has been certified by any secular scientific body, and plausible natural explanations can readily be found for the several on this list for which sufficient documentation exists (Randi 1988).

The Psychic Healers

A recent variation of faith healing called "psychic healing" has had a similar sorry recent history. In the Philippines in particular, charlatans perform "psychic surgery" on simple, trusting, and desperate people. Randi has shown that psychic surgery is a sleight-of-hand trick, with fake blood and animal organs pulled from the body of the patient. He has duplicated bogus psychic surgery on national TV.

The Freidreich Plan

The occasional reported successes of spiritual healing results, at least partially, from human gullibility. To illustrate how the illusion of faith or psychic healing works, Randi has described the "Freidreich Experimental Plan," which he attributes to Dr. Emil J. Freidreich.

The Freidreich plan assures that any remedy, whether it be a drug, psychological or medical treatment, or mystical therapy, will always prove effective for virtually any patient and any disease. Basically, the plan takes advantage of the natural ups and downs of sickness. When the patient feels better after any treatment, the healer claims credit for success. When the patient experiences no improvement, the healer simply finds an excuse for the failure, usually by invoking the "mysteries of God."

Indisputably, people experience feelings of well-being after the "laying on of the hands" of a faith healer, psychic surgeon, or even a conventional physician. Look how a mother's kiss can cure a child's minor scratches. But why assume that this is the result of the intercession of any supernatural forces?

We can understand what happens without invoking the hypothesis of transcendent power. As psychologists Zusne and Jones have explained it, "The kinds of functions controlled by the autonomic nervous system show that a very large number of disorders that are miraculously cured by faith are precisely the disorders that depend on these autonomic functions" (Zusne and Jones 1982).

Miraculous Cures

People have always sought to free themselves from death and the other unhappy inevitabilities of human existence through promises of immortality. They do this despite the complete lack of evidence that these promises are fulfilled. So we should not be surprised that faith and psychic healers flourish, exploiting the occasional healing success that results by chance or psychosomatic means. Witnesses can always be found to testify that a healing method cured them, though proving it is another matter.

For conventional medical cures we have evidence accumulated by careful clinical and laboratory study, and usually some understanding of the mechanisms involved at the chemical and biological level. We can often see the offending organism being destroyed by the prescribed drug under a microscope. If not all conventional cures meet the same standard, this does not detract from the ones that can be demonstrated to be effective. Medicine is an art as well as a science.

The issue is not whether one can dredge up examples of medical procedures not meeting scientific standards, or eventually turning out to be ineffective. Many marginal and folk treatments are used in conventional medicine because physicians are convinced, based on experience, that they are beneficial—or at least are not harmful. When a person is sick, why not try anything that might help?

Rather, the issue is whether *any* of the schemes that violate established scientific principles work. Here I am not talking about harmless folk remedies that ease a person's suffering, but so-called miracles.

No miraculous cure has ever met the standards of scientific method. Until one does, no claim can be made that a miracle has been scientifically demonstrated. People can't call on science to rubber-stamp their beliefs based on their testimony alone. Scientific method requires that an event should not be interpreted as a new phenomenon until all normal explanations can be ruled out to such a level of certainty that the new interpretation is the most probable, that is, least miraculous, alternative.

When the Incurable Are Cured

Many physicians and countless laymen intuitively feel that positive thinking methods, such as Cousins's "Laughter Therapy," help speed recovery. But this has not been convincingly documented to be anything miraculous.

For example, in cases where some kind of time-consuming or painful therapy is involved, the patient's strong desire for recovery makes him or her work that much harder at the therapy. Certainly, patients surrounded by loving, caring friends and family will be kept more comfortable, sleep better, be more likely to take medication on schedule, and have their survival chances improved in any number of small but cumulative ways. But when someone survives a usually fatal illness, nothing forces us to conclude that a miracle has occurred.

One of the aspects of skepticism that proponents of the paranormal constantly bewail as unfair is that it is much easier to debunk a claim than to prove it. While proving a miracle is exceedingly difficult, showing that no miracle occurred is often rather easy.

By applying Occam's razor, skeptics need not even explain the phenomeon observed, just show that all natural explanations are not ruled out. That is, we simply ask whether the data *require* an additional hypothesis, such as the supernatural, to explain them.

Some physicians have kept records of their patients' mental state, religious practice and prayer, use of meditation, and similar factors. Careful studies based on such records rather than subjective impressions have shown no significant correlations that cannot be understood naturally between these factors and the success of treatments for illnesses.

Dr. Saul Silverman, a cancer specialist, has kept careful records on the approximately 6,000 patients he has treated over twenty years. Of the many who were regarded as hopeless by conventional medical standards, twelve patients experienced what could be termed "miraculous cures." That is, these patients, according to Dr. Silverman's conventional medical diagnoses, had "absolutely NO chance of surviving, but they did." They showed no traces of disease five to twenty years after the initial diagnosis. Examining the case histories, Dr. Silverman found no common features such as use of "miracle" drugs, visits to clinics in Mexico, faith or psychic healing, prayer, meditation, or even positive thinking.

Dr. Silverman tells how patients typically react when told they have a probably fatal cancer. They are upset and depressed at first, but many soon convince themselves that they will defeat the disease by faith or will power. Their determination is admirable and undoubtedly helps them through the suffering that follows. But Dr. Silverman found that those who adopt this attitude still eventually weaken and die in the same proportion as those who remain depressed and give up all hope.

One supposedly incurable patient had no will power at all, completely resigned himself to death, but then his cancer totally disappeared. After six years, he finally accepted the fact that he would live after all (Silverman 1988).

The point is that some small fraction of people, perhaps on the order of one in a thousand, survive what medically seems to be a fatal illness. Though the survivors are few in comparison with the many who do not survive, they still constitute many people each year, making room for almost every conceivable "explanation" of the cures. But nothing about these cures forces us to attribute them to supernatural causes. Anecdotal stories of individuals who survive after praying, visiting a shrine, going to a faith healer, or any other such mumbo-jumbo must be measured against the number who do not survive following the same procedures, or survive without any such nonmedical aids.

For reasons not yet fully understood to medical science, some people's immune systems are able to defeat a normally fatal disease. Obviously, elements of good care, proper diet, and exercise, perhaps prompted by a desire to survive, can help and should not be discouraged. After all, what has the patient got to lose? But, the typical patient with a normally fatal disease will die, independent of his or her efforts to invoke supernatural forces.

Many best-selling books on nonmedical healing systems present case study examples taken to be evidence for the validity of the suggested methods. Usually the authors themselves had personal experiences of "miraculous" recovery, and can't believe it was simply an accident. But this kind of anecdotal evidence is not scientific, especially when, as usually happens, the authors select examples that prove their theses, while ignoring many other cases that do not. The database in such instances is inadequate, making any conclusion problematical.

As we will see later when we talk about other kinds of so-called spiritual or psychic experiences, thousands of reports of apparently miraculous cures can be found, but none carries any weight unless it is so well documented that other explanations can be ruled out to a high degree of certitude. As the reader surely can confirm, stories about individual experiences can be greatly embellished in the telling. People are simply poor witnesses, especially when they themselves are the subjects.

We will never find evidence for the paranormal in personal tales and anecdotes. In the case of a miraculous cure, we must have independent, disinterested corroboration. As Abraham Lincoln reportedly said, "The only person who is a worse liar than a faith healer is his patient."

Walking on Fire

The mind-over-matter debate is not about whether the mind or brain can affect the body in unusual ways, but whether it can enable that body to break the laws of nature. Nothing about stigmata—as they really occur, not as they are romanticized—or the recovery from a serious illness, violates any physical or biological principles. Mind plays a role, but not necessarily a paranormal one.

Later I will be discussing other examples where the mind is said to have the ability to violate natural law—in moving objects (psychokinesis) or in producing superhuman strength, as in hypnosis. Here I would like to add another example that more directly illustrates how a misunderstanding of simple physical concepts can mislead people into thinking that something beyond physics is taking place.

A fad that had a mercifully short life in the mid-1980s provides my example—firewalking. Of course, walking on burning coals is an ancient rite of many cultures. But the practice was more recently repackaged in that now familiar American entrepreneurial product, the human potential seminar. Such seminars follow a standard pattern, differing only in the particular gimmick used to provide the special impetus to self-improvement and consciousness-raising. Firewalking was one such gimmick.

Advertisements for firewalking seminars promised the elimination of phobias and addictions, cure of impotence, or mental problems such as chronic depression. For a fee of about $125, you could participate in an organized firewalking experience (Leikind and McCarthy 1985).

In the typical format, the seminars began with an hour of the pop psychology that is associated with the whole genre of self-help seminars. The eighty or so participants were told to believe in themselves, to think positively, to tap into unused capacities both mental and physical: "We only use 10 percent of our brains." The seminars have adopted what became the guiding principle of the New Age: anything is possible if you really want it badly enough.

The unspoken implication was that this included even superhuman feats. The proof was provided by the grand finale: the demonstration of mind over matter in having everyone walk on fire.

While the pit of coals was prepared, the minds of the participants were primed for the event. The excitement built as the coals began to flame. Finally, the participants took turns running across the 1,500°F to 1,800°F glowing fire. Most did so successfully, although some occasionally suffered a mild burn. Very rarely did anyone suffer a severe injury. And when someone did, he or she was left to believe it was his or her fault for lacking the discipline to develop an adequate mental state for protection from the fire. Like the uncured victims at faith-healing circuses, these losers went home worse off than before, both physically and mentally.

Although the seminar leaders were careful not to say so explicitly, the clear implication was that a successful firewalk was the result of the mental state of the participant, that a true mind-over-matter event had taken place. Since over ninety percent of the participants succeeded, they went away happy, feeling that their money had been well-spent and believing that they would derive a permanent benefit from the experience.

Whether any permanent benefit occurs has not been demonstrated. I can cite an example from personal knowledge that indicates otherwise, with statistics to back it up. A few years ago, the University of Hawaii football team was enrolled by their coach in a firewalking seminar. They then proceeded to have their worst season in recent years. They now have a new head coach.

Firewalk Physics

How is it possible to touch your feet to 1,500°F hot coals and not be burned? Doesn't this violate principles of physics? Isn't this paranormal? Yes, it is possible. No, it does not violate any physical principles. No, it is not paranormal.

Imagine a similar situation that occurs when you put your hand inside a hot oven. If you touch the side of the oven or a pan inside, you will get a severe burn. But, if you just briefly place your hand inside without touching anything, you will feel the heat but not be burned.

The reason is the difference between temperature and heat. Heat is the *total* kinetic energy of the molecules in a body. High heat can cause burns because the total energy is high. Temperature, on the other hand, is a measure of the *average* kinetic energy of the molecules. The temperature can be high while the heat is low. In the case of putting your hand in the oven, the air inside the oven has a high temperature, but there are insufficient molecules touching your hand to transmit the heat needed to burn it. We say that air has a low heat capacity.

The hot coals of a firewalk are made of very porous material that becomes even softer and more porous after it burns throughout and turns almost to ash. Ordinary charcoal barbeque brickettes can be used. Their heat capacity is low, and if you move quickly across the coals with wet feet, which provides an additional layer of insulation, you should not get burned. The coal bed must be very carefully prepared and examined to make sure it contains no high heat capacity objects such as rocks or metal. However, it is still possible to get burned trying the trick. Readers are cautioned not to attempt to walk on coals unless they know precisely what they are doing.

I can personally testify that walking on coals requires no special paranormal ability or protection. I have performed the trick myself, and witnessed many others perform it (see Figure 5.1). I must confess that I did get some mild blisters from the experience, but I know others who have walked coals many times without any ill effect—and without a positive thought in their heads.

The Mystical Experience

From these and other examples we will run into in the course of this book, no evidence can be found that the human mind can induce violations of the laws of nature. But what about other powers of the mind? Does the mind possess channels of insight or communication that do not ride on the electromagnetic waves of light or the mechanical waves of sound?

Figure 5.1. The author walking on hot coals.

In both East and West, one of the most convincing arguments for the supernatural is the wide variety of psychological phenomena that are called mystical or religious experiences. Technically, the term mystical experience applies only to the specific feeling of unity or oneness with God, the ultimate reality, as reported by traditional mystics. The true mystic is convinced that this oneness has indeed occurred, and often changes his or her life as the result of it.

According to William J. Wainwright, mystical experiences involve cognitive claims about extramental reality. Mystics believe they have experienced something beyond their own bodies. That is, they are convinced that their experiences were not simply inside their own heads, but rather involved interactions with a transcendental world beyond normal sensory experience.

These experiences are supposedly distinguishable from normal religious feelings such as may happen in church or during prayer. The mystic reports a feeling of wonder and mystery, as well as visions, voices, and intuitive answers to ultimate questions. Furthermore, the ordinary space and time distinctions between bodies is absent; images are shapeless, formless, and the mystic sees things as they "really are" (Wainwright 1981).

I have little doubt that the mystical experience is real. I am sure that many people over the ages have honestly reported what they believed were transcendent revelations. The argument is not over the reality of unusual experiences, but the validity of interpretations of experience that go beyond the normal and the natural.

An analogous situation exists in the interpretation of UFOs. A small percentage of reported strange aerial phenomena cannot be adequately explained because of insufficient data. Some people conclude from this that extraterrestrials are visiting earth in flying saucers. However, the more economical hypothesis is that these are natural effects, hoaxes, or delusions; especially since such explanations can be made for all UFO reports for which there are sufficient data.

Similarly, simply because a mystic claims to have experienced oneness, and proceeds to give up all his earthly belongings to live a life of poverty and meditation, you cannot conclude that he has truly been in contact with a supernatural ultimate reality. His experience still could have been an illusion or delusion. After all, giving up everything is a rather wacko thing to do. The best explanation for wacko behavior is that a person really is wacko.

Some independent corroboration is needed beyond the testimony of the individual. As I have already noted, this is not an impossible requirement. Let a mystic tell us something about the universe that he and no one else knows, and then let that be confirmed by further developments, and we can begin to take mysticism seriously. This has never happened, despite the fact that thousands of people have claimed such revelations.

Mystical experiences go back to the dreams and thoughts of prehistoric people. In *The Origin of Consciousness in the Breakdown of the Bicameral Mind,* Julian Jaynes even goes so far as to suggest that the gods reported in ancient texts to have spoken to early humans were really there—in the heads of these humans. In Jaynes's scenario, the communication between the two halves of the brain was once less effective than it is now, and so ancient people heard voices that sounded to them as if someone else were talking to them (Jaynes 1976).

Voices, Visions, and Dreams

Whether or not Jaynes's theory is correct, humans do hear voices and see visions. For most of us, these occur during sleep or those transition periods between sleep and wakefulness. We have already seen how the dreams of ancient people were probably the original source of the ideas of spirits and gods. Even today, the unreal, aethereal quality of dreams causes people to attribute some source to them beyond their own minds: God, occult forces, astral planes, extrasensory perception.

In our heads while we are asleep, thoughts rattle around in various random and generally meaningless combinations. People try to read much into dreams. The interpretation of dreams was one of the cornerstones of Freudian psychiatry, and huge volumes of data on dreams have been collected over many years. Yet no verifiable, nonchance correlations between dreams and events beyond the knowledge of the individual have ever been demonstrated. With each of us dreaming thousands of dreams in a lifetime, the chances are good that some individual will have a dream that correlates with a significant later event, such as the death of a loved one. We often hear reports of such correlations, but the fact that these occasionally happen can be readily attributed to random chance.

Of course, many people who have these experiences will be convinced that a paranormal event has occurred, and find it easy to believe that they are blessed with superhuman powers. Some may even take up careers as commercial psychics. Few of us need much scientific evidence to conclude that we are individually special. We find it hard to believe that something highly unusual that happens to us personally is simply the result of blind, random chance.

Yet objectively, we must accept chance as the best explanation, without any need to introduce extraneous causal forces for which no independent evidence exists. Remember that billions of people live on this planet, and every day someone, somewhere, will experience a random event whose chance of occurring was less than one in a billion.

The Sacred Disease

Beyond the normal experience of visions and voices in our heads during dreams, a less common experience of visions and voices occurs during periods of wakefulness or illness. And since such events are uncommon, we again tend to attribute deeper significance to them.

This is particularly the case when the event is associated with the frightening convulsions of an epileptic seizure. Some of the most famous people in history have probably had epilepsy: Plato, Buddha, Julius Caeser, Saint Paul, Muhammed, and Napoleon, to name a few (Beyerstein 1988a). Perhaps even Jesus was an epileptic. The voices and visions these prominent people experienced must have influenced their behavior, and consequently greatly influenced the course of history.

Hippocrates (c. 460–370 B.C.E.) called epilepsy "the sacred disease," but he recognized it for what it is: a brain disease. Nevertheless, over the centuries superstitious people continued to regard epileptics as possessed by either gods or demons. Today we can observe on an electroencephalogram (EEG) the abrupt change in brain wave patterns that occurs during a seizure, and treat epilepsy quite successfully with drugs or surgery. Once again, the paranormal has succumbed to natural explanation and scientific control.

We can also explain the "aura" that is observed just before an epileptic seizure begins. In some people, this can even occur without the following seizure. As Beyerstein describes it: "Contact with the environment is not yet lost, but it seems distant and unreal. Visual, auditory, somatic and olfactory hallucinations interweave with veridical percepts. New surroundings may seem peculiarly familiar (deja vu) or frequented places may seem novel (jamais vu). Auras are often fearsome and foreboding, but they can also be ecstatic. Divine, diabolical, or natural interpretations variously appeal to different personalities and social/educational backgrounds" (Beyerstein 1988a).

The dreamy epileptic aura has now been produced by stimulation of the temporal lobes of the brain during neurosurgery, providing convincing evidence that the voices and visions of the epileptic are all in his head.

We can easily see why epileptic visions have been assigned paranormal significance over the centuries. Perhaps the epilepsy of the famous people I have listed helped them gain prominence; people took them to be singled out by God. Isn't that ironic? Has the human race throughout history allowed itself to be enormously influenced, indeed dominated, by people with brain disease?

Knowledge from Seizures?

Other features of epileptic hallucination are also found in mystical experiences. While the individual claims a great feeling of unity and oneness with ultimate reality, the actual visions are very much controlled by his or her personal situation. Voices usually speak in the language the person understands. Christians hallucinate Christ or the Virgin Mary, Buddhists hallucinate Buddha, Hindus see and hear Krishna. No one hallucinates anything that is not already part of his or her experience or imagination.

No record exists of anyone hallucinating about neutrinos or airplanes or computers prior to the twentieth century. No one can point to any new knowledge about the universe first gained during an epileptic vision.

Modern day reports of extraterrestrial UFO "abductions" are very suggestive of temporal lobe epilepsy. Temporal lobe instability in nonepileptics is found to be correlated with reports of dramatic mystical, religious, or paranormal experience (Beyerstein 1988b). In the spirit of Occam's razor, we can say we have a suitable explanation for these reports that does not require invoking any hypotheses other than those relevant to the science of neurophysiology.

Migraines

Far more common than epilepsy is the migraine headache. Migraines affect five to ten percent of the population. I have been a sufferer myself, though in a rather mild form. Although I personally have never had the experience, most severe migraine sufferers report auras and enter trancelike states prior to the attack.

Like epilepsy, migraine is a brain phenomenon that is now well understood. Spasms in cerebral arteries can be triggered by a number of causes: allergy, diet, and stress. This constricts the blood flow to a portion of the brain, starving that region of oxygen and producing the aura experience. The pain comes later, as the returning blood flow stretches the artery walls and pounds the surrounding tissue.

As with epilepsy, some experience the aura without a devastating attack. These people are probably a relatively high portion of the population, which could explain why so many believe they have personally had paranormal experiences. Quoting Beyerstein, "Contributions of these spontaneous warps in the fabric of consciousness to the mystical world-view are probably considerable" (Beyerstein 1988a).

Meditation and Prayer

Not all mystical experiences can be attributed to physical brain dysfunction. Evidently it is possible to use concentration methods to achieve a mental state in which such experiences can occur. The Eastern mystic calls it meditation, while the Western mystic calls it prayer.

The effects of meditation and prayer are operationally indistinguishable. Both serve to empty the mind of its wordly concerns and move it into a state of calm that is then subjectively interpreted as otherworldly. Moreover, after severe fasting, perhaps some temporary brain damage results which then provides the type of aura seen in epileptic or migraine seizures. Because the migraine aura results from oxygen deprivation of a region of the brain, it would seem reasonable that a similar result can be induced by fasting, or possibly psychosomatically.

Collective hallucinations occasionally happen during a highly charged emotional situation. This may have been the case at Fatima, where in 1917 the Virgin Mary is said to have appeared several times to three shepherd children. A similar situation occurred more recently in Yugoslavia. In Texas, hundreds of people claim to have witnessed apparitions of Jesus and Mary during an outdoor mass. These claims are currently being investigated by a panel of bishops.

Illusions and Delusions

Newspapers regularly report the image of Christ seen on the sides of oil storage tanks, burnt tortillas, or similar prosaic objects. Of course, people see what they want to see. It's like the children's game of looking for images in clouds.

Many examples can be cited of people misinterpreting perfectly natural effects, or making causal connections where none exists. In his book *Breakthrough,* Konstantin Raudive claims that he heard "voices of the dead" in tapes he made of the noise on unused radio channels (Raudive 1971). Other people have claimed to see "vitality globules" related to the "etheric body" when looking at people. These apparitions have been explained physiologically as the images formed by red blood corpuscles inside the eye (Zusne and Jones 1982).

In 1964, Jerome Bruner and M. C. Potter experimentally demonstrated the unreliability of human perception. They showed defocused slides to three different groups of subjects. Each group was started off with a different level of blur, but then the slides were gradually focused to the same level of blur. The group that was subjected to the worst blurred original images made the most errors in identifying the images when they came in sharper focus. As Bruner and Potter explained it, all three groups formed hypotheses about the images as soon as they were flashed, and then stuck to those hypotheses even when the evidence indicated otherwise (Bruner and Potter 1964).

In our law courts, we find the common presumption that eyewitness testimony is superior to physical or circumstantial evidence. But the value of eyewitness testimony is dubious indeed, and many cases of mistaken identification by witnesses occur yearly (Wells 1984). Psychological experiments amply illustrate that the data stored in our brains are operated on by our subjective expectations both on the way into the brain and on the way out.

The image of an event stored in our memory is not like the bit pattern on a computer disk. It is more incomplete. But when we picture the event in our mind, we fill in the gaps with our own subjective ideas about how the picture should appear. The ability to do this is a useful quality of the brain, greatly increasing the number of images we are able to store and use. But the price we pay is a less accurate reproduction of the original image when it is recalled. Therefore, we must consider, when we look to human testimony, in law or in science, that we subjectively operate on our images.

Drugs and "Separate Realities"

We have seen that epileptic seizures and migraine headaches can induce visions by cutting off the flow of oxygen to portions of the brain. Hallucinations, produced in this purely physical way, may have played a significant role in religious history.

For example, the drug ergot occurs in a fungus that attacks wheat, rye, and other cereals. It has the effect of constricting blood vessels and reducing the flow of blood. Ergot and its derivatives also have psychedelic capabilities. LSD is a related compound.

Medieval midwives used rotting rye potions to induce abortions. Ergot may be responsible for the hallucinations that caused many women to be accused of witchcraft.

Historians have suggested that periods of wet weather, during which fungi grow more readily, may have resulted in large numbers of people having ergot-induced visions. One example is the rise of the Cult of the Virgin in late eleventh-century France after her appearence to many in visions (James 1986). Chartres cathedral was built on the site where the Virgin is said to have appeared.

According to a recent newspaper report, the outbreak of mass panic and bizarre behavior in France in 1789, considered a pivotal event in the history of the French Revolution, called *Le Grand Peur,* has been attributed to ergot poisoning by University of Maryland historian Mary Kilbourne Matossian.

The prescription drug I take for migraine is called ergotamine and is of the same family as ergot. It acts as a vasoconstrictor.

Today no one would deny that drugs are capable of producing visions. The question is whether the visions signify anything beyond the simple scrambling of signals in the brain by physical or chemical trauma. Some have claimed that drugs open pathways to new realities.

One of the best-selling gurus of the 1970s was Carlos Castenada, who wrote a series of fanciful books about a Mexican peasant magician named Don Juan and the "separate realities" that he alleged were attained through drug-induced hallucinations (Castenada 1968, 1971, 1972). Later, it was convincingly demonstrated that Castenada had written a brilliant piece of fiction (De Mille 1978, 1980, Hines 1988).

Claims of special insights by persons under the effects of LSD were also widely promoted during the decade of turmoil from the mid-sixties to mid-seventies. Scientific experiments were even conducted on the "altered states of consciousness" said to be produced by drugs (Tart 1969).

Convinced by the effects of LSD that they could fly or perform other superhuman feats, any number of people killed themselves by jumping out of windows or off buildings. This was yet another empirical disproof of the notion that mind has the power to repeal the laws of physics. Most researchers today agree that the profound insights of the "drug trip" are a delusion brought about by the drug-induced chemical trauma to the nervous system.

None of these remarks should be taken as a dismissal of the very real value of drug therapy in the treatment of mental disorders, or the carefully controlled use of hallucinogens in brain research. And I readily admit that certain drugs, such as caffeine, may temporarily enhance intellectual powers just as others, such as alcohol, can depress the activity of certain portions of the brain, allowing other portions to dominate. All of this helps prove my point: that thinking is physical.

I merely dispute that the drug trip provides entrance to a spiritual realm. Contradicting the claim that hallucinogens provide a pathway to other worlds is the very fact that they induce hallucinations, which is evidence in support of the physical nature of mind. The brain is no less subject to physical law than the heart, lungs, or stomach.

If thinking were supernatural, then it should be insensitive to drugs or other physical effects such as fatigue, aging, disease, or death. But the evidence is clear and unequivocal: thinking is affected by the same types of physical traumas that affect all other body functions.

Out-of-Body Experiences

With drug-induced trips to the astral plane now out of fashion, new chemical-free attempts to reach beyond the material world have appeared on the scene. The out-of-body experience, or OBE in parapsychological jargon, is one of the latest forms. Popular magazines even give instructions on how to have your own OBE. Anybody can do it. While in a very relaxed or sensory-deprived state, people say they can feel themselves floating out of their body. Often they hover overhead and can see their physical body below.

Parapsychologists have performed laboratory tests of OBE with no result that we can take as serious evidence for anything beyond another form of hallucination, something totally internal to the brain. The OBE particularly lends itself to a rather good test of the psi hypothesis. For example, if someone whose "spirit" or "astral body" is really floating above his or her physical body could read something placed in a position that would be invisible to the physical body, then we might have a good candidate for William James's White Crow of the paranormal.

Parapsychologist Charles Tart claims that this is exactly what happened for one of his subjects. He reported that a young woman was able to correctly read a five-digit number placed on a shelf above her head during an OBE. But she did this only once, in the last of four nights of testing (Tart 1968).

Tart could not demonstrate that the controls were sufficient to have prevented the woman from obtaining the numbers by normal sensory means. In particular, it is suggested that she could have read the numbers reflected in the shiny surface of a wall clock (Zusne and Jones 1982). Since this had not been prevented, for example by simply covering up the clock, we must accept it as more likely than the paranormal explanation. Further, after more than twenty years this result has failed to be replicated.

Near-Death Experiences

The near-death experience (NDE) is another phenomenon extensively reported in the media. This remarkable experience of returning from within a hairbreadth of death is supposedly made possible by the ability of modern medical science to bring people "back to life" after body functions such as respiration and circulation have stopped.

However, similar experiences have been reported throughout history without the benefit of medicine. Many patients who have "returned from the dead" have reported similar experiences of moving through a tunnel toward a bright light—presumably heaven—at the other end (Blackmore 1988).

Actually these people were not really dead. Their brains remained active during the whole period, as evidenced by the waves recorded on an EEG, whenever one happened to be taken at the time.

Death is no longer defined as the stopping of the heart and breathing. One possible operational definition of death is a flat EEG, known as "brain death." To my knowledge, no one has ever survived a flat EEG to talk about that experience. And if the EEG shows brain waves, then the brain is still working.

As with epilepsy and migraines, simple oxygen deprivation of certain sections of the brain can easily explain NDE hallucinations. Visions of heaven are not surprising in the case of persons who are gravely ill or injured. Once again, a good test of the supernatural would be whether insights gained during the NDE experience later turned out to be verifiable.

Past-Life Regression

In the 1960s, a Finnish psychiatrist by the name of Reima Kampman hypnotized a young innkeeper's daughter, named Dorothy, and asked her to "go back to an age before your birth." The psychiatrist and other observers were astonished when she began singing a song in what sounded like Middle English. The girl claimed she had never heard the words or music before.

Dr. Kampman did not rush to write a book about the event, which must have taken some self-discipline. Morey Bernstein's 1952 book on hypnotic regression, *The Search for Bridey Murphy,* had been, after all, a best seller. Instead Dr. Kampman did what had not occurred to Bernstein or other practitioners of the use of hypnotism to help people recall previous lives.

While Dorothy was under hypnosis, the psychiatrist asked the girl where she might have heard the song. In the relaxed state, she remembered browsing through a library at the age of thirteen, and casually picking up a book that happened to be a Finnish translation of the *History of Music* by Benjamin Britten and Imogen Holst. There she found the song "Summer Is Icumen In" with the words rendered in a simplified medieval English (Harris 1986; also see Harris in Basil 1988, p. 130).

In the case of Bridey Murphy, newspaper reporters for the *Chicago American* demonstrated that not all the media are easy marks for phony paranormal claims. These reporters were able to show that Ruth Simmons (really Virginia Tighe), the subject of the Bridey Murphy case, could easily have obtained the knowledge of Ireland that so impressed everyone. Her relatives had visited model Irish villages, complete with replicas of the tower of Blarney Castle, at the 1893 World's Columbian Exposition in Chicago and the St. Louis Fair of 1894. Of course, they would have talked about what they had seen at family gatherings (Harris 1986). But if Bernstein had asked Virginia where she learned so much about Ireland, and related the likely truth in his book, would he then have had a best seller?

Seeing the Future

Throughout history, and undoubtedly before history, people who seemed able to predict the future have been especially revered for their evident possession of paranormal powers. However, looking back, we can find no prediction that meets adequate criteria to be declared paranormal.

The first criterion would be that the prediction must be specific. The oracle at Delphi was particularly noted for stating her predictions in vague phrasing that could be interpreted many ways. In a commonly cited example, Croesus, the king of Lydia, asked if he should go to war with Persia. The oracle told him that if he crossed the Halys river between Lydia and Persia, he would destroy a great empire. Encouraged, Croesus invaded Persia and was defeated. Returning to Delphi and complaining to the oracle, he was told that he should have paid for another question: Whose empire would be destroyed—Persia's or his own? The oracle could not lose, regardless of whether Croesus won or lost.

Modern psychics and astrologers use the tricks of ancient soothsayers, and many new ones as well, to fool the gullible. One psychic in Hawaii predicted in December of 1986 that Halley's comet would not strike the earth after it made its turn around the sun. I suppose every night he predicts that the sun will rise the next morning.

People often ask me about Nostradamus, whose prophecies in the sixteenth century have been the subject of a number of popular books. In a TV show starring Orson Welles, the writings of Nostradamus were interpreted to predict a California earthquake that was to occur in May 1988. The quake did not occur as predicted. I think old Welles pulled another hoax on us, this time from the grave. Recall that it was Welles who terrified the population with his realistic radio broadcast in 1938 of H. G. Wells's *War of the Worlds.*

Nostradamus's predictions are all vague and easily interpreted to suit any occurrences. I especially like Woody Allen's parody of the phrasing of one of Nostradamus's predictions: "Two nations will go to war, and one will win." Nostradamus himself would probably have been highly amused.

To be credible, predictions must be specific. But they also must be risky, not just statements of the obvious or highly likely. Even the eventual occurrence of a risky prediction must be evaluated with caution. Phony seers will usually intersperse their numerous obvious and safe predictions with a few unlikely ones, on the chance that they might make an accidental, spectacular hit. These unlikely predictions usually fail miserably, but the seers understand that people usually forget the misses but will notice the one lucky hit.

At the beginning of each year, newspapers and magazines are filled with the predictions of psychics and astrologers. In a recent healthy development, articles are appearing that also list the predictions made the previous year, noting how many have not come to pass, showing that psychics do no better than the rest of us in foretelling what really lies ahead.

Precognitive Dreams

Of course, given enough chances for something to happen, it probably will, and thousands of predicted unlikely events have been reported in the media and occult literature. Much of what is presented as evidence for paranormal phenomena is of this unreliable, anecdotal nature.

The most common form of anecdote from personal testimony is the precognitive dream. This usually involves dreaming that a loved one has been injured or killed, and then finding it to be true. We can't tell how much these stories are embellished, but even if they are not, they do not strike me as very strange.

Parents are always worrying about their children, especially during those dangerous teen years when the youngsters are full of energy, rebelliousness, and inexperience. Given all the parents with teenager and post-teenager offspring speeding along the nation's highways at three o'clock in the morning, is it really surprising when an occasional parent happens to have a nightmare which comes true? To calculate the odds of this occurring by chance, though, one would need to know how many times the nightmares do not come true.

The night after writing this I happened to dream that my long-time friend and colleague in the next office had died. I was relieved, but not surprised, to see him alive and kicking the next day. We each had a good chuckle over the incident, but I did not write it up and send to to the *Journal of the American Society for Psychical Research*. If you examine that journal and others that report claimed psychic events, you will find no record of dreams that did not come true, or studies on the number of times Jesus' face has failed to appear on a burnt tortilla.

Coincidence

People are always surprised by what they think are unlikely coincidences. However, many such coincidences are really not so unlikely. For example, in a room of twenty-five people, it would seem intuitively unlikely that two would find they have the same birthdate. After all, there are 365 days in the year, so the odds for a particular pair to have the same birthday are a low 1/365. But there are 300 different ways to pair twenty-five people, so the odds are almost even (300/365) that one pair will have the same birthdate. Nothing psychic about it.

I heard of a story told by a fellow physicist. Working in a foreign country, he would drive out each morning to the nearby airport to buy a U.S. newspaper. One night he dreamed that he was at the airport and watched in horror as a green-and-white airplane rolled down the runway and crashed. After awakening the next morning and making his usual airport stop, he was startled to see a green-and-white plane that he had not noticed before. With a great feeling of dread he watched the plane taxi down the runway, take off, and safely disappear into the distance.

References

Basil, Robert. 1988. *Not Necessarily the New Age*. Buffalo, N.Y.: Prometheus Books.

Beyerstein, Barry L. 1988a. "Neuropathology and the Legacy of Spiritual Possession." *Skeptical Inquirer* 12-3:248.

———. 1988b. "The Brain and Consciousness: Implications for Psi Phenomena." *Skeptical Inquirer* 12-2:163.

Blackmore, Susan. 1988. "Out of the Body?" in Basil 1988. p. 165.

Bruner, J. S., and Potter, M. C. 1964. "Interference in Visual Recognition." *Science* 144:424.

Butler, Kurt, and Rayner, Lynn. 1985. *The Best Medicine*. New York: Harper & Row.

Castenada, Carlos. 1968. *The Teachings of Don Juan: A Yaqui Way of Knowledge*. Berkeley, Calif.: University of California Press.

———. 1971. *A Separate Reality: Further Conversations with Don Juan*. New York: Simon and Schuster.

———. 1972. *Journey to Ixtlan: The Lessons of Don Juan*. New York: Simon and Schuster.

Cousins, Norman. 1976. "Anatomy of an Illness (as Perceived by the Patient)." *New England Journal of Medicine* 295:1458.

De Mille, R. 1978. *Castenada's Journey*. 2d rev. ed. Santa Barbara, Calif.: Capra Press.

———. 1980. *The Don Juan Papers: Further Castenada Controversies*. Santa Barbara, Calif.: Ross-Erickson Publishers.

Harris, Melvin. 1986. "Are 'Past-Life' Regressions Evidence of Reincarnation?" *Free Inquiry* 6:18.

James, John, 1986. *The Traveller's Key to Medieval France*. New York: Alfred A. Knopf.

Jaynes, J. 1976. *The Origin of Consiousness in the Breakdown of the Bicameral Mind.* Boston: Houghton Mifflin.

Leikind, Bernard J., and McCarthy, William J. 1985. "An Investigation of Fire-walking." *Skeptical Inquirer* 10:23.

Randi, James. 1986. *Flim-Flam!* Buffalo, N.Y.: Prometheus Books.

———. 1988. *The Faith Healers.* Buffalo, N.Y.: Prometheus Books.

Raudive, K. 1971. *Breakthrough.* New York: Taplinger.

Silverman, Saul. 1988. "Miracles Happen—But Beware of the Proclaimed Causes." (Private communication).

Tart, Charles T. 1968. "A Psychological Study of Out-of-the-Body Experiences in a Selected Subject." *Journal of the Americal Society for Psychical Research* 62:3.

———. 1969. *Altered States of Consciousness.* New York: John Wiley & Sons.

Wainwright, William J. 1981. *Mysticism.* Madison, Wisc: University of Wisconsin Press.

Wells, Gary L., and Loftus, Elizabeth F. 1984. *Eyewitness Testimony.* Cambridge: Cambridge University Press.

Zusne, Leonard, and Jones, Warren H. 1982. *Anomalistic Psychology: A Study of Extraordinary Phenomena of Behavior and Experience.* Hillsdale, N.J.: Lawrence Erlbaum Associates.

6.

On the Matter of Mind

The chimpanzee and the human share about 99.5 percent of their evolutionary history, yet most human thinkers regard the chimp as a malformed, irrelevant oddity, while seeing themselves as stepping stones to the Almighty.

Robert L. Trivers in the Foreword to
The Selfish Gene by Richard Dawkins

Cogito Ergo Sum

They say that war makes a person think. Not having been to war myself, I can't speak from personal experience. But judging from the stories I hear, most people placed in harm's way think about their mortality, the apparent meaninglessness of life, and other matters of a deep, but personal and subjective nature. But when Rene Descartes (1596–1650) fought in the Thirty Years War in 1619, he thought about mathematics. Holed up to escape the cold on a stormy November day in Bavaria, he heard thunder, saw lightning, and received the insight of analytical geometry.

The Cartesian coordinate system, in which points are represented in terms of their projections x, y, and z on three mutually perpendicular axes, remains the most commonly used method for specifying the position of a body in space. Students are taught it as a prelude to their first halting steps in physics.

Descartes believed that the physical universe could be understood in terms of mechanical laws, and described mathematically and geometrically. In his particular model, natural movement was circular, with planets and other bodies enmeshed in whirling vortices of atoms. However, his model was vague and lacking in predictive power. Furthermore, Johannes Kepler (1571–1630) showed that the planets follow elliptical rather than circular paths.

A generation after Descartes, Isaac Newton developed precise mechanical laws that successfully described the motion of bodies. He showed that the rules of elliptical planetary orbits, which had been discoverd by Kepler, logically followed, according to the mathematical formulae, from the laws of motion and gravity. In this way, the idea of the mechanical universe was successfully implemented.

Although Descartes helped trigger the scientific and technological explosion that has so dominated history since his time, he was nevertheless a product of that time. He still found it necessary to regard God and the "rational soul" as entities transcending the mechanical universe. Descartes could not imagine how mechanical particles and forces could think and reason, and so he maintained the duality of mind and matter that had been remembered from antiquity.

Descartes divided all reality into three parts: God, matter, and the individual rational soul, which he equated to mind. According to the traditional view of materialists, matter is the most evident reality. The equally traditional view of theologians holds that God is the primary basis for philosophical discourse. Descartes disagreed with both views, taking the existence of the rational soul as the starting point from an initial position of doubt.

Rene Descartes was part of the Age of Reason that had a revolutionary impact on all phases of human thinking, from the philosophical to the political. The key principle of this age was that all authority is open to question. Following that principle, Descartes said that we can doubt everything but the very existence of doubt itself. God and matter may be illusion, but the thoughts within one's own head must be real. In Descartes's famous phrase, "*cogito ergo sum* (I think, therefore I am)" (Descartes, *Discours de la methode,* 1629).

Pure Thought

In Descartes's view, matter and God become known to us only through mind. Their realities are not inferred from sensory data, from empirical facts. Rather, they are innately known by virtue of their own essence. The idea of intrinsic, a priori, knowledge would be seconded in the following century by Immanuel Kant (1724–1804).

Both Descartes and Kant were greatly impressed that the theorems of Euclidean geometry could be derived from a few simple axioms. This seemed to be an example of knowledge about the real world that could be determined by reason alone, without reference to any observations about the world. Reason thus joined revelation as a source of immutable truth about natural events. Reason was considered an innner, mental source.

Much later, mathematicians showed that Euclidean geometry, taken by both Descartes and Kant as descriptive of the inherent mathematical structure of the physical world and, by inference, according to Kant, a demonstration of the power of pure thought, was simply one of an infinite number of possible geometries that could be applied. In 1915, Einstein would use non-Euclidean geometry to describe the motion of massive bodies in his general theory of relativity.

Because of the success of the general relativity theory, we regard space as being basically non-Euclidean. In the face of this fact, the pure-thought inference of a Euclidean universe is erroneous. And so is the pure-thought inference of any geometry. The geometry of the universe is, like everything else about the universe, a matter of observation. And so, the example of Euclidean geometry used by Kant to demonstrate the power of pure thought establishes quite the opposite. It turns on its head and provides another example of the final authority of sensory data on all matters concerning the material world.

The Ontological Proof

Descartes's reasoning from *cogito ergo sum* to the existence of God was even more spurious. Basically, he revived the centuries-old "ontological proof": that which exists is more perfect than that which does not; thus, a perfect being must exist if we can conceive of such a being. Descartes needed to place God somewhere in the picture to keep the Church happy. He saw what happened to Galileo just a decade earlier, when the Church forced the great Florentine physicist and astronomer to recant his teaching that the earth moved around the sun.

Once Descartes had "proven" the existence of God, however, he solved another problem: how to prove the reality of the material world without recourse to observation. If only the content of pure disembodied thought cannot be doubted, then everything else, including the material world, could simply be an illusion. Descartes resolved this conundrum by saying that God, being perfect, would not deceive us with sensory impressions that were not real.

Even Descartes's contemporaries could see the flaws in his reasoning. The philosopher Thomas Hobbes (1599–1679) reportedly said that Descartes should have stuck to geometry. The physicist Christiaan Huygens (1629–1695) thought that the French philosopher had woven a romance out of a metaphysical web. Neither believed that Descartes had really demonstrated the separation of matter and mind (Durant and Durant 1961).

The Ghost in the Machine

If mind and matter are separate substances, then how do they interact? In explaining how the nonmaterial mind interacts with the material body, Descartes could think of nothing original, falling on the usual rationalization given by those who cannot explain the acts of God, such as letting the innocent suffer while the bad prosper: He has his own mysterious ways.

In the Cartesian dualist picture, the human body is composed of matter that obeys the laws of mechanics. However, in addition to the levers, electrical wiring, and chemical factories of our bodies, we possess a nonmaterial soul that performs the thinking and reasoning processes. The soul is eternal, transcending space and time, with powers not limited by the laws of mechanics. It can perform miracles. The soul is the ghost in the machine.

Still we seem to carry our souls around with us as we move in space and time. The soul must interact with the material world someplace within our bodies. Descartes located the point of that interaction in the pineal gland at the base of the brain, since this was the only organ in the brain that did not seem to be duplicated elsewhere in the brain. There could, after all, be only one soul.

As we will see, however, four separate sections seem to exist in the human brain capable of highly independent action. People whose communication lines between their left and right cerebral hemispheres have been severed literally have two separate minds. So if the soul is mind, in the usual dualist equation, we have a real problem with multiple minds. Which mind survives the body? Which of our multiple minds is the one that we associate with the individual personality?

A Sense of Humor

As we will find, the Cartesian duality of mind and matter is not supported by the evidence. Mind can be understood as an exclusive property of matter. But first, let us take a look at what modern philosophers, the descendents of Descartes, have made of the mind-body problem.

Philosophers of science perform the useful task of observing science from the outside and trying to determine what principles and practices work successfully. In the past, they have concentrated mainly on the physical and biological sciences. However, more recently, philosophers have turned their attention to psychology and its attempt to explain human behavior.

Psychology is a comparatively new science that is still searching for clarity in its terms and methods. Further, it must contend with the wide-ranging conceptions and misconceptions of common views of mind and behavior, pop psychology, and folk psychology. Many of the terms used in psychological jargon originated from traditional folk views.

Familiar terms, however, are not necessarily scientifically meaningful. Using words in a grammatically correct sentence does not automatically guarantee that the concepts they describe are logical. One can form para-doxical sentences, such as "This sentence is wrong."

Not all folk concepts are meaningful. For example, it was once thought that heat was a substance. This substance was identified by the word "caloric." However, when physicists discovered that heat was not a substance, the word caloric disappeared from the lexicon.

Similarly, in medieval days it was thought that the body contained four fluids: blood, phlem, choler, and humor. Perhaps the state of mind called "humorous" will someday be rendered as obsolete as the body fluid once called humor. And, if the evidence continues to build that mind is a product of matter, perhaps many of the mental terms we associate with the traditional spiritual elements of mind will also disappear into oblivion.

A Category Mistake

Gilbert Ryle, in *The Concept of Mind,* says the "dogma of the ghost in the machine . . . is entirely false, and false not in detail but in principle. It is not merely an assemblage of particular mistakes. It is one big mistake of a special kind. It is, namely, a category-mistake" (Ryle 1959).

To Ryle, the mind-body problem is a pseudo-problem. That is, it really isn't a problem, though we have made it into one by our own fuzzy thinking and sloppy use of language. In his view, called "behaviorism," mental events are not of the same logical type as physical objects. And these two different logical types relate to two different ways of describing observed human behavior.

In the behaviorist view, when we use terms that indicate mental char-acteristics, like intelligent, prudent, virtuous, stupid, hypocritical, and cow-ardly, to describe a person, we are simply describing some general features of a person's actions. Another class of terms could be used that relate to sensible phenomena. If the two classes closely correspond, then they are essentially equivalent.

For example, an intelligent action might be indicated by a numerically high score on a university test or IQ exam, or writing a best-selling book. A virtuous action might be donating time and money to a worthy cause or sacrificing one's life for one's country. In each of these examples, a single physical event has taken place that can be described one way or another, depending on the speaker's choice, either as a factual report (donating money) or mental act (of virtue). It's like snapping pictures of an object from different angles; the object remains the same.

Introspection

However, to most people, feelings and thoughts are as real as the objects of their senses. It would appear futile to try to convince someone that her joy, pain, or grief are not every bit as real as the chair she sits in. She *knows* they are real.

In philosophy, this is called the argument from introspection. It basically assumes that we have a faculty of inner observation that matches our faculty for outer observation. Many are convinced that this inner faculty can penetrate closer to the essense of reality than our senses and the neural apparatus that processes their data.

But, even if this inner faculty exists, consider how untrustworthy it actually can be. It is capable of telling us anything we want to hear, of providing us with the wildest fantasies. People walk around thinking they're Napoleon or Jesus Christ. Victims of certain brain lesions can be totally blind, yet stumble around convinced that they can see everything around them. Even if introspection were a means for obtaining knowledge, how could we distinguish that knowledge from fantasy except by testing it against sensory observations?

Mental States

Regardless of our positions on the validity of mental states, we should all be able to agree that an important qualitative difference exists between the two types of phenomena we call mental and physical. The objects of the physical world are observed by many people, who can then reach a consensus on their size, shape, color, and other characteristics. We can usefully analyze these objects with respect to commonly agreed-upon categories.

Assuming they represent a valid concept, mental states are confined to a single individual's private, subjective thoughts. They cannot be defined in the same objective way that we define physical quantities, such as the distance between two points or the mass of a body.

This means that internal mental states cannot be studied with the same methodology that science applies to external events, which heavily relies upon objective consensus and empirical replication. This leads to two possibilities for the scientific study of mind: either we admit that mental states are not per se useful concepts, and recast them in terms of objective, verifiable and replicable external phenomena; or we must develop a methodology that uses the unverifiable concepts of internal mental states.

Of course, it has been the task of psychology since Freud to develop a methodology to deal with mental phenomena. Most psychologists have generally assumed the basic validity of terms used to describe mental states and activities, both those carried over from folklore to newly-invented ones such as "ego" and "id." However, any limited success the inner-state psychological enterprise has gained can largely be attributed to the couch-side manner of the therapist—art rather than science. By contrast, the more conventional materialistic neurosciences have enjoyed immense success, including the chemical treatment of many mental diseases that traditionally have been attributed to spiritual causes.

Suppose we were to discard our traditional concepts of inner mental states and try to relate the phenomena they represent in terms of external observations to which objective criteria can be applied. Then mental-state terms simply become categories for classifying certain general types of behavior, without filling in the details of the precise phenomenal actions. And if mental-state terms describe events in the sensory world, analogous to terms used in physics like mass and energy, then they are part of the sensory world and not some "world of inner space."

Words and Thoughts

When we have thoughts that we wish to communicate to the rest of the world, we are required to put those thoughts into words and images that originated in sense impressions about that world, which is outside our heads. We cannot identify disembodied thoughts, independent of the physical world, that are present in a separate entity of mind.

Philosophers have come to realize that our understanding of concepts is tied very closely to language. Ludwig Wittgenstein has argued that language structures human thought, and even what we take as introspective knowledge, like Descartes's *cogito ergo sum,* is inherently the result of a formulation using words (Wittgenstein 1953). Numerous studies have demonstrated how language determines or affects thinking in different cultures.

Words do not exist as entities in the mind independent of the real world of sensory data, since they relate directly to that world. They originated as particular sound wave patterns, and have since been translated into written form and computer bits. Words are not matter, but they would not exist without matter. So if some of our mental constructs are formulated with words, then they are structured according to the world of matter.

The psychologist Theodore Sarbin has given a nice example of how the development of a word for a mundane physical phenomenon has led us to attribute some profound internal condition to what is really a sense-world observation. The word "anxiety" originated from the thirteenth century French word *anguisse* for pain in the throat. It came to be used metaphorically (like pain in the neck?) to refer to a type of feeling. Forgetting this meta-phorical origin, we now assume that there exists a mental state called anxiety, just because we have come to use a word to describe a certain human feeling or type of behavior (Sarbin 1968).

Theories of Mind

Philosophers dealing with the mind-body problem have identified a number of theories on the nature of mind. These include several different forms of dualism and materialism: substance dualism, property dualism, reductive materialism (also known as the identity theory), and eliminative materialism. Other theories not necessarily dualist or materialist include functionalism and behaviorism, although both of these are consistent with materialism. (For good nonspecialist introductions to theories of mind, see Churchland 1979 and Fodor 1981.)

At first glance, the "identity" theory of mind seems to follow most directly from the common understanding of materialism. In the identity theory, mental states are directly equated to physical states of the brain. But, as we have seen, mental phenomena are of a different nature than physical phenomena: subjective rather than objective, inner rather than outer. Thus identifying an exact, one-to-one correspondence between mental and physical states seems an unlikely prospect.

Eliminative materialism is another materialist option. As Churchland summarizes this viewpoint, "the one-to-one match-ups will not be found, and our common-sense psychological framework will not enjoy intertheo-retic reduction, *because our common-sense psychological framework is a false and radically misleading conception of the causes of human behavior and the nature of cognitive activity.* On this view, folk psychology is not just an incomplete representation of our inner natures; it is an outright misrepresentation of our internal states and activities" (Churchland 1979, p. 43). Eliminative materialism is thus even more extreme than the identity theory, suggesting that our everyday mental concepts do not have any scientific value.

Functionalism provides a third materialist alternative, one that views mental states, including at least some of those carried over from folk psy-chology, as valid categories of reality. For example, pain is a real experi-ence that cannot simply be reduced to a neurobiological effect, nor eliminated from rational consideration of human psychology. Functionalists use the computer metaphor, likening human mental states to the software of a computer.

The various categories of theories of mind that have been identified by philosophers allow considerable room for a purely materialist interpretation of mental processes. In particular, note that the materialist view does not stand or fall on the ultimate success of reductionism. Even if we cannot uniquely relate mental states to specific physical states of the nervous system, we still can conclude that mind is a product of matter. Let us now examine the evidence that supports this conclusion.

Brain and Behavior

Until this century, the belief in the supernatural nature of mind led people to attribute nonphysical causes to every type of human behavior. A person acting virtuously was guided by "spiritual inner light," by angels or higher deities. A person behaving in violent or antisocial ways was "possessed by the devil."

Today a large body of scientific evidence supports material rather than spiritual forces as the source of behavior, good or bad. Behavior can be traced to the brain, at least as a plausible hypothesis, not requiring the additional hypothesis of any supernatural quality.

Beyerstein has listed the evidence that behavior and the physical processes in the brain are correlated: (1) there exists an evolutionary relationship between brain complexity and the cognitive attributes and intelligence of animal species; (2) mental abilities increase as an individual, and his or her brain, matures; (3) mental functions can be diminished or lost as the result of brain injury, illness, or aging; (4) many mental operations can be correlated with physical changes in the brain; (5) drugs affect many modes of thinking (Beyerstein 1988).

On top of all this, we have the remarkable results of split-brain studies, in which different forms of cognition, language, and analytical and synthetical reasoning have been located in different hemispheres of the brain and observed to operate independently when the communication lines between these portions are severed. We will discuss these further below.

Brain Waves

Various mental states regarded by many as spiritual in nature can be correlated to measurable physical properties of the brain. The evidence supporting these assertions is widespread, and I need only touch on an illustrative portion of it.

First, the brain emits low levels of electromagnetic radiation: brain waves. Electromagnetic radiation is produced by the electrochemical currents in the brain that carry signals back and forth between the basic brain units, the neurons. This radiation is tiny, undetectable a few meters away by even the most sensitive receivers. However, by means of simple electrodes placed on the skull, an EEG of brain wave patterns can be made and correlated with various mental activities.

When the brain is busy with important mental tasks, no easily identifiable harmonic components in the brain wave patterns can be recognized, and the EEG trace appears as random "white noise." But, as the brain relaxes, alpha waves with a frequency in the range 8 to 13 Hz (1 Hz or Hertz = 1 cycle per second) first appear, then gradually become replaced by slower delta waves (4 to 8 Hz), and finally by theta waves (0.5 to 3 Hz). The white noise observed during high brain activity is characteristic of an electrical system with many components in simultaneous and largely independent action. This effect is analogous to a large number of tuning forks of different pitches struck randomly, or a large orchestra with its members individually tuning up before a performance.

Rest and sleep are generally accompanied by synchronized harmonic waves, while wakefulness is characterized by noise, occasionally interspersed with periods of alpha and other synchronized waves.

The conclusion that thinking is a physical process is strongly supported by the observation of noisy brain waves during periods of active thought. Again, you can argue that this does not "prove" the nonexistence of a spiritual component; I will again simply counter with Occam's razor: nothing in the data forces us to introduce the additional hypothesis of such a component in human thinking.

Positron Emission Tomography

Much newer technology than the EEG is providing researchers with far more powerful tools for probing the structure of the brain. One example is Positron Emission Tomography (PET). In a PET scan, blood or glucose is given a radioactive tag and injected into the bloodstream of the patient, who is then observed performing various functions such as seeing, hearing, speaking, or thinking. If a particular region of the brain is primarily responsible for that activity, then blood will be shunted to that region to provide the energy needed.

The radioactive elements used in PET are short-lived positron emitters produced artificially in a cyclotron accelerator near the site. Positrons are positively charged antielectrons. When a nucleus of the radioactive element decays in the brain at the brain site to which it has been carried by the blood flow, the positron from the nucleus quickly annihilates with any nearby electron, producing two gamma ray photons. These gamma rays, each with an energy of a half million electron-volts, easily escape from the skull and can be detected by instruments placed around the patient's head. The signals are then fed to a computer that can produce beautifully detailed pictures of regional brain glucose uptake, highlighting the brain areas associated with various mental tasks (for some excellent color examples, see Montgomery 1989).

Do Computers Think?

Rene Descartes's original argument for the supernatural nature of mind was based on his inability to conceive of a machine being able to think. His viewpoint is certainly understandable, given the limited level of technology in his day. However, we now have machines that show many of the characteristics of thinking: computers.

Despite the unpopularity of the idea that our brains are basically some form of machine, developments in computer science point squarely in that direction. To those who have argued that computers must be programmed and thus cannot duplicate human thinking, A. M. Turing has given a ready answer: humans are programmed too—by their genes (Turing 1950). And, as we will see, machines can model other aspects of human thinking besides the logical and arithmetic operations we normally associate with computers.

Most researchers in computer intelligence, or artificial intelligence (AI), say they can detect no quality of logical processes that cannot be reproduced, at least in principle, with a machine. The amazing progress in this field has happened after only a few decades of computer development and experience. What can we expect in another century? I predict that one day computers will exhibit new modes of thinking that the human brain cannot accomplish.

For the brain developed, after all, by the largely random processes of natural selection and survival of the fittest. How much superior can machines become, developed as they are by the rational processes of science? And, if it can be shown that humans possess mental capabilities exceeding those of current primitive computers, who's to say that within a few more generations computers will not possess such capabilities, and much more?

Now the position I have staked out above is not without its detractors. These have no trouble finding adherents in the general community, given the common intuition that the human mind cannot simply be a computer. Basically, their argument relies on the fact that mental acts do not resemble those of computers. The mind seems to possess capabilities that cannot be reduced to computational algorithms.

Turing had shown that a machine can always be built able to solve any problem which is reducible to computational mathematics. As far back as 1956 it was demonstrated that a computer can solve problems in symbolic logic. So if thinking is just logic and computation, the brain is simply a machine. Conversely, if we can prove that all thinking processes are not simple computation, then some think we will have shown that the brain is more than a computer.

Perhaps the most sophisticated recent argument relies on the deep nature of mathematics (Penrose 1989). It was once thought that mathematics is simply a form of logic. However, it is now known that mathematical truths exist that cannot be proven. In 1931 the Austrian logician Kurt Gödel proved that many mathematical systems, including arithmetic, cannot be made internally logically consistent. Thus mathematical truths exist that cannot be derived by computational algorithms. If the mind is able to grasp these truths, it must do more than compute algorithms.

The claim is basically that the brain can think thoughts that computers can't, because computers are forced to be logical. That is, the human mind is often illogical—so it can't be a computer! Actually, I have trivialized the argument a bit. A computer with an open wire might give an illogical answer. Similarly, illogical human behavior could simply result from poor wiring in the brain. But that is not what is being argued. The claim is that a fully functional human brain uses nonlogical thought processes in developing mathematical theories that, by their universality and further development by others, ultimately turn out to be correct. Somehow the mind, in a Platonic and Kantian way, can determine truth without sensory input after all.

Even so, this doesn't mean that the brain cannot still be material, that all thinking including the nonlogical cannot be accomplished by known physical processes. Can we build a machine that acts nonlogically? Of course! Can we build a machine that solves problems by other than sequential computer algorithms? Yes! In fact, such machines are patterned after the processes that we believe take place in the brain.

Neural Networks

Just as AI research has helped us to put human intelligence into perspective, neuroscience is now beginning to have an impact on AI. The suggestion has been made that current bottlenecks in electronic computing might be solved by methods that are actually modeled after the human nervous system.

These new methods use neural networks, in which large numbers of gates, or neurons, are interconnected with each other in such a way that the strengths of these connections can be automatically varied. The network then learns to solve problems without the kind of detailed direct programmer input that is required in conventional computing.

Neural networks have now been used on a number of pattern recognition problems that defied solution by previous techniques. Because of their unique human-like quality, neural network systems have been called "cognizers" instead of computers. The word is derived from "cognition," which is the mental process by which knowledge is acquired, through reason, perception, or intuition (Johnson and Brown 1988).

Cognizers exhibit many of the features we would associate with non-logical thinking. For example, they obtain solutions that are often not the most optimum. And their solutions often cannot be reduced to computational algorithms. So in artificial neural networks, we seem to have a viable material model of the human nervous system that answers the objections of those who see qualities in human thinking that cannot be accomplished by algorithmic computers.

The Brain as Computer/Cognizer

So if the brain is a machine, we must be careful not to draw too closely the analogy between it and the modern computer. At the very minimum, detailed differences in what is called the "architecture" exist between brain and computer. I will later explore the general question of whether purely material systems are capable of exhibiting the properties we associate with mind. Nothing requires that systems with these abilities look anything like today's computers. The neural network machines, which more closely emulate the human nervous system, are still in their infancy.

Current computers have localized memories and central processing units (CPUs), while the nervous system distributes these functions among its various portions, including those outside the brain proper. This was a consequence of the largely chance processes of evolution. Moreover, the survival imperative of evolution provided redundancy: operations normally performed by one portion of the brain are carried out elsewhere when that portion is injured or destroyed. Computers might usefully emulate this feature, since they still break down far more often than the human brain. The existence of this cranial redundancy has been amply demonstrated in cases of people with brain injuries, or where surgery has destroyed a specific region of the brain. New connections then form in another, healthy portion.

These differences in architecture between artificial machines and the brain are expected to become less pronounced as computers develop further and cognizing systems based on neural networks find a practical niche.

So whatever differences now exist should not discourage us from seeing the underlying, in principle similarites. Brains, artificial and natural, are organized matter. (For a recent airing of the debate, see Searle 1990 and Churchland and Churchland 1990.)

The Reticular Activating System

If the brain has no specific CPU, a kind of control unit seems to be located in the brainstem: the reticular activating system (RAS). The signals from this system to the rest of the brain are required to keep various regions of the brain awake. The RAS acts as a regulator, deciding what will be given attention. In this way, data processing bottlenecks are avoided and the brain's resources can be used for the most important tasks at hand. Like the operating system of a computer, the RAS sets priorities for brain activity.

What algorithm does the RAS use in setting its priorities? Basically, events are classified into expected and unexpected. The brain handles expected events with little thought, reacting to them in a way conditioned by previous experience. The unexpected requires more processing power to decide on a response.

Orienting and Habituation

When a surprising stimulus occurs, a number of body changes take place. The sensitivity of sense organs increases, blood is pumped to critical areas, and glands increase their production of certain chemicals. While all this is happening, a noisy, asynchronous EEG is observed indicating activity of several portions of the brain. This burst of brain activity is called the "orienting" reaction, as the body adapts or orients itself to the new situation.

As the experience becomes familiar, the "habituation" process—habit forming—takes over. A copy of the brain's original response to a stimulus is recorded and saved for use when similar stimuli happen again. If a stimulus changes its features, the orienting process again takes place (Furst 1979).

Evolution of the Brain

Even simple forms of life, such as the jellyfish, utilize the same fundamental electrochemical mechanisms as we do for communicating information throughout their bodies. Our system has as its basic unit the nerve cell, or neuron. Animals with neural systems "behave"; that is, they react to various stimuli. More primitive animals, like the sponge, do not react, or do so very minimally. While a jellyfish can swim to where nutrients are, a sponge must lie in one place and rely on the ocean current to bring it food. So a neural system provides greatly added survival value (Thompson 1985).

The Reptilian Brain

As more complex organisms evolved, their nervous systems also naturally evolved to increasing levels of complexity and capability. We can see these different levels of development in the human brain itself. Deep inside the human brain today is the small thalamus that, together with the brainstem, is a relic of the brain of our reptile ancestors.

The old reptilian brain is programmed for repetitive, stereotyped behavior: searching for food, mating, migrating (MacLean 1970). Within this system is the important hypothalamus that is linked strongly with sex and aggression. It is also the center of punishment, reward, and pain.

As mammals developed along a separate evolutionary branch from reptiles, the reptilian brain of the mammal gradually was surrounded by what ultimately became the limbic system in higher mammals and humans. In humans, stimulation of the limbic system can produce hallucinations. Schizophrenic hallucinations involve chemical imbalances in the limbic system. This very primitive portion of our brain is probably responsible for many of the religious "visions" that have so affected history.

The Smell Brain

The earliest mammals had to find a way to survive during the millions of years when the giant reptiles ruled the earth. This they accomplished by operating at night, and since their eyes and ears were insufficient as sources of data about the external world, mammals developed a powerful sense of smell. Unlike the fairly straightforward data from sight and sound, however, smells are complicated and ambiguous, and require additional, more intelligent, processing. The circuits mammals developed for this added task gradually evolved into the cerebral cortex that is now the seat of our higher intelligence (Jastrow 1981).

Why Are We So Much Smarter Than Animals?

The human brain is really not much different in size and structure from that of the chimpanzee. Yet we have far greater intellectual capabilities. This is a puzzle for which there are a number of possible solutions. Perhaps the human brain exceeds a certain critical mass that is required for higher intelligence. However, intelligence is not strictly correlated to brain mass. Whales and dolphins have larger brains and are presumably less intelligent than humans. Also, there is no correlation between brain size and intelligence within the human community.

It was once thought that our prominent forebrain, the frontal lobes, contained the intelligence that set us apart from low-browed apes. But this is no longer believed to be the case, though the forebrain certainly has important functions. Our high intelligence is more likely a matter of cerebral cortex area, with all the folds and wrap-arounds of the cerebral gray matter resulting in maximum surface area.

Of course another explanation for the great superiority of humans over animals strikes at the theme of this book: is the human being set apart from the rest of the animal kingdom, by special supernatural powers? Even before we begin the search for supernatural powers of the human mind, we need to ask whether the powers the human mind exhibits are truly unique.

Animal Language

Many characteristics of human thinking are evident in animals: rudimentary language, toolmaking, foresight, and insight. Let me just say a few words about language. Many animal species communicate by sounds, but this is not what we conventionally mean by language. Human language involves much more than commands and reactions to these commands, and animals do not seem to have developed such further capabilities naturally. Nevertheless, humans have now successfully taught two exceptional species, chimpanzees and dolphins, rudimentary language.

Chimps are now able to assemble simple sentences and create new words. For this they have been taught to use a sign language, since chimps do not possess the vocal mechanisms to produce human language sounds. Significant results that indicate real language processing in the chimpanzee brain have been achieved. Combining signs to form new expressions not in his taught vocabulary, the chimp Washoe gave a watermelon the name "drink-fruit." Washoe even invented a sentence of his own, saying "more tickle" when he wanted to play a familiar game (Gardner and Gardner 1969). No one expects chimps to ever exceed the language ability of a three-year-old child, and not everyone is convinced that the results so far are significant, but it appears likely that this crucial element of intelligence, the ability to communicate and indeed think symbolically, is not the unique preserve of the human brain. Besides, computers can do it.

Split Brains

Fascinating insight into the operation of the brain has come from split-brain studies, pioneered by Roger Sperry, who received the Nobel Prize in 1981 for this remarkable work.

In a very radical treatment for severe epilepsy, communication lines between the two halves of the cerebral cortex have been severed. Such operations appeared at first to be a great success; the violent epileptic seizures that the patients suffered were eliminated or greatly relieved, without any obvious reductions in mental capacity. After recovery, the patients functioned more or less normally. However, interesting side effects began to appear when patients were tested under special experimental conditions.

Sperry and others experimented with split-brain patients using what are called lateralized inputs. In one of the first such experiments, the patient put her hand behind a curtain, and a key was placed in the hand. Later she was able to pick the key from a group of objects when she used the same hand. However, she was unable to select the correct object if asked to use the other hand (Furst 1979).

As a general though not exclusive rule, data from the sensory apparatus on the left side of the body go to the right cerebral hemisphere and vice versa. A split-brain patient acts as though she has two separate "minds" that cannot communicate effectively with one another. For most right-handed people, the left brain performs language processing. A split-brain patient cannot vocalize what is felt in the left hand, since the data from that hand pass to the right hemisphere, which is largely incapable of language.

In the normal person in such a case, the right brain would pass the processed information to the left, which would then vocalize it. The split-brain patient can use her left hand, however, to point out objects she cannot vocalize, allowing us to conclude that the right hemisphere still has significant intellectual capability.

Many similar experiments in which visual data from the left and right portions of a visual field are used as input have confirmed these findings. Although popular expositions of split-brain studies have tended to over-simplify the complex patterns observed in these experiments, a number of crude generalizations can be made.

For the typical right-handed person, the left hemisphere is associated with analytic thinking. It analyzes data sequentially, piece by piece, assigning words and symbols to each piece. The right hemisphere is associated with visual-spatial thinking—what is called synthetic thinking. It puts the pieces of data together into a whole, forming global concepts. Unlike the popular misconception in which the left brain is considered the seat of rational intelligence and the right brain the seat of artistic creativity, both sides are intelligent and creative. As we will see, both are needed in solving complex problems that require a combination of analysis and synthesis.

People with damaged left hemispheres cannot speak words. But, incredibly, they can sing these same words if part of a song! People with damaged right hemispheres have trouble dressing themselves or finding their way around their own neighborhoods. Split-brain patients have trouble associating names with faces, and drawing with their right hands.

Two Brains or More?

Apparently, the human brain evolved at least two separate programs to allow for different but equally useful and important forms of thinking. At the risk again of oversimplifying the complete story, the left hemisphere's program is a serial code, sequentially processing elements that change with time, such as speech sounds. The programs of most of today's computers are serial codes.

The brain's right hemisphere program, by contrast, is a parallel code that processes data from all regions of space simultaneously. Today's supercomputers operate with parallel codes that enable them to solve certain problems far faster than serially coded computers. Neural network machines, or cognizers, also operate on the parallel principle.

Logical thinking, which is sequential, thus predominates on the left; intuitive, big picture thinking on the right. The two hemispheres can interact in solving a problem. For example, consider solving a physics problem. A good student first assesses the overall global features of the problem: drawing a picture, labeling it, stating what is given and what is to be found, writing down the general principles that may apply. Then she proceeds to obtain the solution by means of a series of mathematical steps. Finally she assesses the answer, and asks if it makes sense and is consistent with the original global overview.

Sperry suggests that a duality of consciousness may exist even in normal people (Sperry 1961). We seem to have at least two brains and maybe more. MacLean has proposed that we have three brains (MacLean 1970).

My University of Hawaii colleague neurochemist Bruce Morton has gone McLean one better and proposed a model he calls the "quadrimental brain," in which he associates the four Freudian constructs with the four localized brain elements, as follows:

Morton's Quadrimental Brain

Id	=	Reptilian Brain (Stem-Cerebellar Striatal System)
Ego	=	Limbic System
Superego	=	Right Cerebral Hemisphere
Intellect	=	Left Cerebral Hemisphere

So What is the Soul?

Now if we still have souls, as most people believe, then what portion of the brain do we associate with that soul? Conventionally, the soul is one with our individual personality, which is supposedly able to survive the disintegration of our material bodies. But every quality that we regard as an aspect of individual personality seems to be already geographically located within our material multiple-brain nervous system.

For example, we are taught that violent, unthinking behavior results from an evil soul. Yet that type of behavior is traced to our ancient reptile brain that controls our sometimes violent reactions to perceived threats. Do reptiles have souls?

Descartes associated the soul with the rational mind, but we have seen that analytic and synthetic reasoning are associated with the two cerebral hemispheres. In the modern form of mysticism referred to as New Age, the right hemisphere is regarded as the seat of spiritual awareness. It is not explained why spirit should be so localized.

And what about higher emotions, more sublime than reptilian rage? Surely love and hate must be part of the soul's qualities. But emotions seem to be seated in the limbic system. Well, you can see the problem. All the normal qualities we associate with human personality, every aspect of human thinking, from raw survival emotions to the highest planes of analytic and synthetic reasoning, are traceable to specific localities in our brains.

Certainly the discovery of the left-right hemisphere split is one more example of the physical nature of the processes we call mental. To turn this discovery around, as some have done, and take it as evidence for a spiritual quality of mind defies both the analytical reason of the left brain and the intuitive reason of the right. It makes no sense on either side of the brain.

Altered States of Consciousness

One further region on the map of experience may be explored for evidence of the soul, or more precisely, for evidence of a nonphysical component to human thinking. The four qualities of brain that I have discussed—analytical and synthetical reasoning, emotion, and autoreaction—may not cover all of human mental capacity. There may be yet other, less common qualities of the mind, so-called altered states of consciousness.

Again we are dealing here with what is, to a great degree, a matter of definition. What we mean by human consciousness is itself difficult to define in the same operational manner as physical quantities such as temperature and mass. Some have even suggested doing away with the concept of consciousness. So we should not be surprised to find little consensus on what constitutes an altered state of consciousness. Nevertheless, there is widespread belief that certain uncommon mental states exist. Furthermore, these altered states of consciousness are supposed to provide selected humans with the power to perform miracles, or to communicate with realms beyond the senses.

Hypnotism

One familiar powerful altered state of consciousness is the hypnotic state. Most people think of the hypnotic trance as an exceptional mental state during which a human is capable of superhuman physical or mental feats. Hypnosis was originally called animal magnetism by its discoverer, Frederick Anton Mesmer (1733?–1815), conjuring up images of magnetic field lines emanating from the bodies of living organisms. However, no evidence for such field lines has ever been found. In a later chapter I will discuss the modern reincarnations of the idea of special fields associated with living things: Kirlian auras and "energies of consciousness."

No special brain wave patterns, apart from those associated with drowsiness, are seen in hypnotized subjects. Hypnosis is simply a highly relaxed and suggestible state (Barber 1970). The issue is not whether hypnotized subjects are capable of feats of strength or able to withstand pain at levels not common in the normal, awake state, but whether such feats go beyond the bounds of natural law, that is, whether they are miraculous and supernatural.

A performing hypnotist will usually call a group of people to the stage and, during the preliminary banter, select out the few he senses will be particularly suggestible. Hypnotic effects can be produced with techniques other than the mumbo-jumbo usually associated with hypnotism. And despite widespread belief to the contrary, no act performed under hypnosis violates any laws of physics. Like so much in the area of extraordinary human capabilities, the story of hypnotism is filled with gross exaggerations that do not hold up under carefully controlled observation.

Meditation

As we saw in the last chapter, meditation is widely believed to be a means by which the mind can access realities that transcend those presented to our senses. Certainly, meditation practice generally involves alterations in perception, feeling, and thought that are sometimes quite profound, or at least considered so in the mind of the practitioner (Tart 1969).

But, as with hypnosis, nothing forces us to conclude that something extraordinary or paranormal is happening, despite a vast literature of claims that meditators achieve higher levels of consciousness.

Transcendental Meditation

A popular form of Yogic meditation taught in the United States is transcendental meditation (TM). It was brought to the West in 1958 by Maharishi Mahesh Yogi and attracted considerable attention, especially after the Beatles and other celebrities became involved in it during the 1960s.

The Maharishi (meaning "great teacher") is a physics graduate of Allahabad University in India who became a monk and studied for thirteen years in the Himalayas. During that time he developed TM, a form of meditation based on ancient Hindu techniques in which the subject uses a repeated sound or "mantra" to empty the mind of other thoughts and induce a highly relaxed state.

TM is far from unique. For example, it is quite similar to the Coué treatment, developed around the turn of the century by the French pharmacist Émile Coué. Coué had his patients repeat the familiar refrain: "Day by day in every way, I am getting better and better." Studies by Harvard cardiologist Herbert Benson have also shown that the same beneficial "relaxation response" can be obtained by repeating "Hail Mary, full of grace," "Lord Jesus Christ," "Our Father who art in heaven," "The Lord is my shepherd," "Shalom," or the word "one." Benson concluded that all major religious traditions use simple, repetitive prayers (Benson 1975, 1984). This could account for much of the healing power claimed for prayer and meditation. Perhaps it works—but that does not make it supernatural. In fact, Benson's data support a purely physical interpretation.

The TM state, according to the Maharishi, opens a channel to "cosmic consciousness." Of course, not just any mantra will suffice; rather, one must be specially assigned by a certified TM instructor. In fact, there are very few different mantras. However, the fee for sufficient TM training sessions to achieve "adequate meditation skills" and a personal mantra can make a significant dent in the pocketbook. This seems to be the strongest physical effect of TM.

In recent years, TM has cultivated a more scientific image. Proponents have published numerous studies showing effects of TM on the body such as lowered blood pressure and decreased respiratory and heart rates. Much of this work is done by Ph.D. scientists at the Maharishi International University in Fairfield, Iowa. The effects they report are probably real, and if so, again tend to confirm the physical nature of mind. Whether the studies have actually demonstrated that body changes are unique to

TM or any other form of prayer or meditation, and not just the result of exceptionally relaxed conditions, is problematical.

I would not mind if TM proponents were simply satisfied to claim that they had developed a useful relaxation technique and quietly pursued scientific studies to understand its nature and properties. But they do not. Instead, the wildest claims are made. For example, during the heyday of grand unified field theories of particle physics a few years ago, TM advertising claimed a vague connection with the grand unified field. When the grand unified field was not found in physics laboratories, the ads disappeared.

Claims have also been made that TM enables people to levitate, obviously breaking the laws of physics. These claims are often accompanied by photographs showing levitators in the lotus position apparently suspended well above the ground. Investigation quickly reveals, however, that the levitators have simply learned how to hop in the air from the lotus position, and the snapshots give the appearance of suspension in mid-air. Perhaps the research at Maharishi International University has some merit, but this type of dishonest and exploitative behavior on the part of the TM organization contributes little to the credibility of that research.

Consciousness

Even if all the mental processes we have discussed are of material origin, or at least provide us with no reason to hypothesize otherwise, there remains a more general mental concept that is continually associated with powers beyond the realm of the senses: consciousness itself.

Is there even a quality or feature than can be defined, in an operational, scientific way, as consciousness? Sir John Eccles, who received the Nobel Prize in 1963 for his work on synaptic mechanisms, has said, "We can, in principle, explain all our input-output performances in terms of the activity of neuronal circuits; and consequently, consciousness seems to be absolutely unnecessary! . . . As neurophysiologists we simply have no use for consciousness in our attempts to explain how the brain works" (Eccles 1966, p. 248).

Although this sounds like the the last word on the subject, Eccles only intended to remove consciousness from the realm of reductionist neuroscience, not from the realm of all observable phenomena. In fact, he is very much a dualist, asserting that a world exists separate from matter that contains all subjective and mental experiences, with a "Divine Providence operating over and above the material happenings of biological evolution" (Eccles 1979, p. 235).

So what is consciousness? Let me beg off attempting a definition of a word with such wide-ranging and complex associations. I'll leave that to philosophers and psychologists. People frequently use the word "consciousness," so I will assume there's something to it, that a set of phenomena exists that can be classified as consciousness. My immediate concern is whether the phenomena we associate with consciousness show any evidence for anything other than physical matter. If consciousness does not exist, then my job is already done.

Like Eccles, Roger Sperry has written voluminously on his personal concept of mind and consciousness. He insists that his view, contrary to Eccles's, is neither supernatural nor dualistic, but is most definitely non-reductionist.

Sperry does not deny the material basis for mind as a function of the brain, but because of the usual pairing of materialism with reductionism, Sperry prefers not to call his ideas materialist; rather he uses the term "mentalist." Putting jargon aside, the substance of Sperry's view is that mind is one of a whole class of superior properties that emerge from material systems when they achieve a certain high level of organization (Sperry 1987).

Sperry argues that, while the role of submicroscopic particles and forces in the structure of brain and mental processes cannot be denied, it is not the whole story. He says that properties of experienced consciousness are "different from events out of which they are built" (Sperry 1969). Rather, they are something more, something beyond the physics of elementary particles. They are emergent properties of organization that have the power to exert "downward control" over the microscopic constituents of the system.

For example, the atoms in an automobile are pushed around under the control of the driver. So, "mind and consciousness are in the driver's seat. . . they give the orders, and they push and haul around the physiological and the physical and chemical processes as much as, or more than the latter processes direct them" (Sperry 1983, p. 31).

Beyond Physics?

Is Sperry's mentalism really inconsistent with the materialist view, as he seems to imply? I don't think so. Materialism does not forbid parts of a complex system from interacting to produce an intricate structure with capabilities the individual parts lack. The computer is a good example. A computer is constructed of circuit boards interconnected by wires. The circuit boards are made of microchips and printed connections. The microchips are made of silicon and other atoms, and operate by the laws of quantum mechanics.

Similarly, artificial neural networks have been built out of material components. The people who design computers or neural networks use no special laws of physics. They use systems design principles, but these still treat the computer as an assemblage of individual components. And the properties that emerge as the result of the computer system of microchips, circuit boards, and wires do not thereby become part of some new realm beyond the world of matter out of which they spring.

The emergence of new properties at every level of organization, from molecules to cells to organisms to brain, does not imply that some new class of nonphysical or superphysical laws exists to govern the behavior of those systems. The laws of biology or neuroscience may appear quite different in form from Maxwell's equations or quantum electrodynamics, but the latter remain behind the basic electrochemistry of cell metabolism or neuronal interactions.

I do not deny the existence of emergent properties of complex physical systems, resulting from their organization. If the downward control over the atoms of those systems that Sperry talks about is simply a case of pushing the atoms around, consistent with the laws of microscopic physics, then so what? Large bodies obey Newtonian classical mechanics as they push around the atoms of their systems. That does not mean that Newtonian mechanics forces atoms to violate quantum mechanics.

The New Consciousness

The issue is whether the downward control of emergent properties is so great as to produce violations of physical laws. While this is not Sperry's claim, his words seem to be interpreted that way by some in the New Age movement, providing them with a holistic paradigm for the new consciousness and its modernized occult world view (for example, see the summer 1988 issue of *Revision: The Journal of Consciousness and Change*).

In later chapters, I will explore in detail the distinction between the materialist reductionist picture of the universe as discrete bits of matter interacting locally, and the holistic view that instantaneous connections exist between all events across space and time, violating current principles of physics. I call the latter view "strong holism," in order to distinguish it from the simple notion that the whole is greater than the sum of the parts. This view, which is also called holism, is a trivial, noncontroversial form; I call it "weak holism."

The important fact to realize about strong holism is that, although it may provide comfort to those who want to believe that they are a part of some great cosmic whole, a mighty price has to be paid for the existence of such a connection: the violation of well-established physical principles such as Einstein's theory of relativity.

The new consciousness is supernatural, and I wish that influential scientific figures like Roger Sperry would realize this. They also need to object more vigorously to the way their ideas are exploited by those who promote occult causes. I, for one, am unwilling to tear down the whole edifice of modern physics, whose worth is proven by its success, without the strongest evidence that it needs to be torn down. Wishful thinking and comforting words about our participation in a "living cosmos" are not enough.

The issue is not simply whether the whole is greater than the sum of its parts. As I have already noted, it certainly can be, as with a crystal of salt, a computer program, or a symphony, without violating any laws of physics. The issue is whether evidence exists for some type of aetheric field, a cosmic consciousness, that somehow links thoughts instantaneously throughout the universe, with the power to violate the physical laws of that universe. In other words, is there a ghost in the machine?

References

Barber, T. X., et al. July 1970. "Who Believes in Hypnosis?" *Psychology Today.*

Benson, Herbert. 1975. *The Relaxation Response.* New York: Morrow.

Benson, Herbert, and Proctor, William 1984. *Beyond the Relaxation Response: How to Harness the Healing Power of Your Personal Beliefs.* New York: Times Books.

Beyerstein, Barry L. Winter 1988. "The Brain and Consciousness: Implications for Psi Phenomena." *Skeptical Inquirer* 12-2:163.

Brown, J. 1977. *Mind, Brain and Consciousness.* New York: Academic Press.

Churchland, Paul M. 1979. *Matter and Consciousness.* Cambridge, Mass.: Bradford Books.

Churchland, Paul M., and Churchland, Patricia Smith. January 1990. "Could a Machine Think?" *Scientific American.* P. 32.

Dawkins, Richard. 1976. *The Selfish Gene.* Oxford: Oxford University Press.

Durant, Will, and Durant, Ariel. 1961. *The Story of Civilization VII: The Age of Reason Begins.* New York: Simon and Schuster.

Eccles, J. C. 1966. Discussion after "Consciousness" by E. D. Adrian. In *Brain and Conscious Experience,* J. C. Eccles, ed. New York: Springer.

———. 1979. *The Human Mystery.* New York: Springer.

Fodor, Jerry A. January 1981. "The Mind-Body Problem." *Scientific American.* P. 114.

Furst, Charles. 1979. *Origins of the Mind.* Englewood Cliffs, N. J.: Prentice-Hall.

Gardner, R. A., and Gardner, B. T. 1969. "Teaching Sign Language to a Chimpanzee." *Science* 165:664.

Jastrow, Robert. 1981. *The Enchanted Loom.* New York: Simon and Schuster.

Johnson, R. Colin, and Brown, Chappell. 1988. *Cognizers: Neural Networks and Machines That Think.* New York: Wiley.

MacLean, P. D. 1970. "The Triune Brain, Emotion and Scientific Bias." In *Neurosciences,* F. O. Schmitt, ed. New York: Rockefeller University Press.

Montgomery, Geoffrey. March 1989. "The Mind in Motion." *Discover.* P. 58.

Pagano, R. R., Rosen, R. M., Stivers, R. M., and Warrenburg, S. 1976. "Sleep During Transcendental Meditation." *Science* 191:308.

Penrose, Roger. 1989. *The Emperor's New Mind: Concerning Computers, Minds and the Laws of Physics.* Oxford: Oxford University Press.

Ryle, Gilbert. 1959. *The Concept of Mind.* New York: Barnes and Noble.

Sarbin, T. R. 1968. "Ontology Recapitulates Philology: The Mythic Nature of Anxiety." *American Psychologist* 23:411.

Searle, John R. January 1990. "Is the Brain's Mind a Computer Program?" *Scientific American.* P. 26.

Sperry, R. W. 1961. "Cerebral Organization and Behavior." *Science* 133: 1749.

———. 1969. "A Modified Concept of Consciousness." *Psychological Review* 76:532.

———. 1983. *Science and Moral Priority.* New York: Columbia University Press.

———. 1987. "Structure and Significance of the Consciousness Revolution." *The Journal of Mind and Behavior* 8:37-88.

Tart, Charles. 1969. *Altered States of Consciousness.* New York: John Wiley & Sons.

Thompson, Richard F. 1985. *The Brain: An Introduction to Neuroscience.* New York: W. H. Freeman.

Turing, A. M. 1950. "Computing Machinery and Intelligence." *Mind* 59:433.

Wallace, R. 1970. "Physiological Effects of Transcendental Meditation." *Science* 167:1751.

Wittgenstein, L. 1953. *Philosophical Investigations.* New York: Macmillan.

7.

Searching for the Spirit

I think that it is about time that the truth of this miserable subject "Spiritualism" should be brought out. It is now widespread all over the world, and unless it is put down soon it will do great evil. I was the first in the field and I have the right to expose it.

From the 1888 confession of Margaret Fox Kane

The Definitions of Words

Like most others who promote the virtues of science to the general public, I am often confronted with the assertion: "Science is your religion!" My flippant answer is: "Religion is your science!" In most such confrontations the argument is of little substance, simply coming down to definition. No effective communication occurs when the words used mean something different to the antagonists in an argument, as is usually the case when the participants adhere to such extreme points of view as religion and science.

So the first thing we must do when we try to communicate is make clear the definitions of our terms. In science, this is very effectively managed. The words and symbols of science are carefully defined, usually by international consensus. The definitions are independent of the language used. Ideally, scientific definitions are operational, that is, they are formulated in terms of some well-prescribed measurement procedure to ensure that everyone would get the same results, within measurement error, when they measure a quantity under the same conditions.

Time and Space

Thus, time is operationally defined as what is measured on a clock. The basic unit of time, the second, was once defined as $(1/60)(1/60)(1/24)$ of a mean solar day. However, in 1967, an international agreement redefined a second as the time required for a cesium-133 atom to undergo 9,192,631,770 vibrations.

Arbitrary? Of course. Time is a human invention, so we humans can define the basic unit anyway we want. This particular choice was the number of cesium vibrations that most closely matched the previous definition and was chosen simply for convenience. Using this definition, or one like it, enables us to use simpler equations to describe events. If we had continued on with the old one, atomic processes all over the universe would have to be gradually speeded up in our theories. The new definition provides a more economical description that corresponds to the slowing down of the earth's rotation under the action of tidal forces.

What about space? Since Einstein showed that the speed of light is a constant, distance can simply be defined as the time it takes light to travel between two points. The meter is operationally the distance traveled by light in a vacuum during $1/299792458$ second.

Notice that both time and space are measured by clocks. Since all other physical variables such as mass, energy, temperature, and magnetic field are defined in terms of space and time, the clock becomes the fundamental measuring device in our descriptions of the physical world.

Seeking a Consensus

The operational definitions of space and time illustrate one of the virtues of science: little argument occurs, or at least should occur, over the meaning of its terms once those terms are agreed upon. Science operates on the basis of a consensus about terms.

That is not to say that scientists never get into arguments. Of course they do, but the consensus-seeking mechanism of science provides the means for the settling of disputes. Scientific arguments usually converge on a consensus, as empirical facts adjudicate a position that the whole scientific community ultimately accepts. Sometimes it may take years, other times this can be accomplished in months.

By contrast, political or religious arguments only converge on a consensus with great difficulty. If they do, it often happens by force of arms or economic power, rather than the merits of arguments. Historically, politics and religion have been marked by schisms, continual splits into warring groups that take irreconcilable sides on issues. Neither group understands the inability of the other to see the unassailable logic of its argument. What the partisans often fail to realize is that both sides of an argument can be equally logical; the disagreement is usually not about the process by which conclusions are drawn, but about the different assumptions that led to those conclusions.

Most often, neither side even recognizes that the starting point of its argument is an assumption rather than a self-evident fact. Just try arguing with fundamentalist Christians sometime. You will be incapable of getting them to even conceive of the notion—much less accept it—that the starting point of all their claims is the *assumption* that the Bible is the literal word of God.

We should make every effort to avoid any misinterpretation of words. But to do so completely, we would have to carefully define each word used. No one, not even a philosopher, does this. Aristotle came close to this ideal, even defining the term definition. Inevitably, however, any writer must assume some common ground of definitions between his reader and himself.

So too here. When I use a scientific term, I mean it as it is operationally defined by the consensus of scientists. And when I have used a word that is not normally a scientific term, then I have taken it to mean what the majority of people who speak the English language take it to mean.

Incidentally, dictionaries are not always of much help in this regard, since they tend to give all the usages of a word, even those which may be rare or inconsistent with current convention, or incompatible with the other usages listed. For example, the recently published second edition of the *Oxford English Dictionary,* the most definitive work on the English language, lists as one of the new definitions of the word *bad,* "excellent." That's what rock star Michael Jackson means when he sings, "I'm bad, I'm bad, I'm really bad."

Defining Science and Religion

Well, is science a religion? Or, is religion another kind of science? Some definitions of religion take it to mean any system of belief, thought, and behavior that attempts to address the most fundamental issues of the origin and nature of the universe and our place as human beings in that universe. If we use this definition, then certainly science is a religion, since science also deals with these issues within a framework of confidence, even faith, in the validity of its methods.

Of course, we could have started by defining science as any system of belief, thought, and behavior that addresses these same fundamental issues, and then conclude that religion is science. But in either case, we would not be applying the meanings that most people accept when they use the terms science and religion. To almost everyone, science and religion are distinct systems of thought, and we would be in violation of that consensus were we to equate the two. Religion and science, by conventional usage, are not equivalent.

Over the centuries, the term religion has been taken by most people to refer to systems of thought in which the answers to ultimate questions are found in a supernatural, transcendent reality not immediately evident to the senses. Knowledge of that reality is thus obtained, not from the senses, but by means of extrasensory channels, particularly the revelations of certain holy men and women who have convinced others that they have been chosen by a deity to possess this channel.

Science, on the other hand, deals exclusively with the world of our senses and seeks to describe the order observed by any individual looking at the same data. This order is expressed in terms of principles that we call natural law. And the key to the acceptance of an interpretation of natural law is consensus.

The principles of order we call natural law are the same for all observers, at all places and times. They do not arbitrarily change at the whim of any individual. A phenomenon is not accepted as scientific until it has been repeatedly seen by independent observers, and a physical principle used to describe these observations must involve risky predictions about other observations. Until these are confirmed, the principle remains, at best, provisional.

Thus, the difference between science and religion is intimately related to the difference between the words "natural" and "supernatural." Science seeks a natural explanation for the events of the universe, while religion claims a supernatural one. Accepting this, we are still left with defining a distinction between natural and supernatural. That's not as easy as you might think. Perhaps the major task of this book is to try to make the distinction clear.

Distinguishing Natural and Supernatural

Following conventional usage, we classify as natural all phenomena that we can detect with our senses. Supernatural phenomena are then not directly accessible to our senses; they exist beyond the senses, although their effects may be sensed. But this does not mean that all nonsensible phenomena are supernatural. For example, the human eye is unable to detect infrared light or X-rays, but these are certainly natural phenomena, easily recorded on special photographic film.

Numerous phenomena such as infrared and ultraviolet light, ultrasound,

X-rays, gamma rays, and neutrinos are undetectable by the human senses but detectable by modern instrumentation. These phenomena are, by any reasonable definition, natural. It would be irrationally egocentric of us to only define as natural those phenomena accessible to the limited sensory apparatus of the human body. Even the sound of a dog whistle would then be supernatural, and dogs would have a version of ESP not possessed by humans.

Beyond this definition of natural as phenomena detectable to our senses, or the instruments we have built to aid these senses, there could also be undetected natural phenomena for which there is not yet sufficiently sensitive detection equipment. If such phenomena have no effect on any but other undetectable phenomena, then we can forget about them. With no detectable effect, why worry about them? They have no influence on anything, and so are for all practical purposes nonexistent.

So the issue is reduced to whether there exist undetectable phenomena that have a detectable effect. Shall we classify these as natural or supernatural? If they have a detectable effect, isn't that precisely what we mean by detectable, and hence natural? Only if a phenomenon were undetectable *in principle* could we safely classify it as supernatural. But then, how could we ever prove that it exists?

Supernatural Impossible to Verify?

So the existence of the supernatural appears to be impossible not only to logically define, but also to verify. However, I will not go so far as to say that the concept should be summarily ruled out. The laws of physics are impossible to verify as well, and philosophers still argue whether they can be unambiguously defined. We cannot verify that the law of conservation of energy will always hold in every situation, but we note that it does hold in every situation examined so far and we use it to build the engines of industry. In other words, we provisionally accept the laws of physics, as long as they continue to perform useful tasks for us. Verification is, to some degree, a matter of practicality; if it works, we use it.

Why not give concepts of the supernatural the same shake? If a phenomenon can be found that dramatically violates a number of the most basic principles of physics—energy conservation, causality, relativity—we might be permitted to make a tentative hypothesis of the existence of transcendent reality beyond the realm of natural phenomena, and then try to see if it works.

This is the only logical way I can imagine for distinguishing natural and supernatural. Natural phenomena obey certain fundamental laws of nature that have been confirmed, but never finally proven, by consistent, long-term observation. Supernatural phenomena would then violate these laws. That's okay, provided that these phenomena can be shown to produce some effects or have some application that is beyond the scope of the natural.

Obviously, disagreement with natural law has to be profound. It would become an acceptable explanation only when all natural explanations, even those far more unlikely than a supernatural explanation, are ruled out. Since the quantum mechanical uncertainty principle allows for the violation of certain physical laws at some well-defined level, supernatural phenomena must go well beyond these limits.

The Natural Paradigm

I will call the scientific world view "the natural paradigm." The natural paradigm is currently implemented at the fundamental level in the form of the Standard Model described earlier. However, the ultimate overthrow of this model, which I regard as inevitable, will by no means overthrow the natural paradigm.

As we have seen, the Standard Model describes the universe as a conglomerate of fundamental bodies that interact with one another through the exchange of other fundamental objects. These exchanges result in bodies moving toward or away from one another. When a large number of bodies move toward one another, they continue this exchange process and aggregate into composite systems that become nuclei, atoms, molecules, rocks, trees, planets, stars, and galaxies.

The behavior of fundamental bodies is orderly to a sufficient extent that this order can be used to predict, on the average, future behavior. The rules of order that describe this behavior are the laws of nature.

However, quantum mechanics holds that the behavior of an individual body cannot always be predicted with certainty; only the average behavior of ensembles of bodies is described by the laws of physics. Individual bodies will often deviate from this average behavior, so that an inherent uncertainty exists in nature. Or in other words, some events happen by chance—undetermined, uncaused, and unpredictable.

The Disintegration of Traditional Beliefs

A century ago science had reached a pinnacle. Newtonian mechanical principles, supplemented by related principles in optics, thermodynamics, continuum mechanics, and electromagnetism, successfully described every observation in the laboratory, with a few exceptions. These few exceptions—

black body radiation, the photoelectric effect, spectral lines of atoms—would be the anomalies that triggered the twin twentieth-century revolutions of relativity and quantum mechanics.

But the remaining structure of nineteenth-century physical science remained vast and compelling. Furthermore, the immense practical value of the physical sciences was becoming evident to most human beings on earth, as technology transformed their lives.

Simultaneously, the biological sciences were making enormous strides. In probably the most revolutionary development of the times and the greatest intellectual event since Newton, Charles Darwin's formulation of the theory of evolution challenged the traditional view held by virtually every major religion and culture: that human beings are creatures in a separate class from animals, residing on a higher plane of existence, some place between earth and heaven.

Darwin had found convincing evidence for the ancient idea that humans are simply another species of animal, naturally evolving from other forms of life, along with every other species. Like the theories of Galileo and Newton, the arguments presented by Darwin were so compelling that they were accepted almost immediately by the majority of the scientists of the day, despite their almost total contradiction of existing religious dogma.

Applying Science to the Supernatural

Within the nineteenth-century milieu of technological and intellectual wonders wrought by science, and the accompanying disintegration of traditional beliefs, the methods of science began to be applied to the world of soul and spirit previously the exclusive preserve of religion and philosophy. Biblical scholars looked at historical and scriptural texts with more open and doubting minds, to try to extract an element of truth about what really happened in the holy lands of Palestine and India thousands of years ago. What they found was not always consistent with traditional teachings. It became clear that not everything in scripture could be true, unless one discarded the whole rational framework of science and replaced it with fantasy.

Scholars also used the new investigative and critical techniques to explore the origins of human culture. They found an immense variety of primitive experience, far removed from the mythology of the Old Testament, in which all humanity is said to be descended from the survivors of the Flood. People brought up as Christians could not help but be jarred by the conflict between the Judaeo-Christian picture and the implication forced by these new developments that their beliefs were founded on superstition, myth, and legend rather than historical fact.

The quest for knowledge of human origins is usually justified by our need to know who we are and where we came from. But a secondary motive also lay behind many of these critical searches into the past. With the new discoveries casting doubt on everything that had been taught during their upbringing at home, church, and school, educated men and women sought ways to reconcile the diverging views.

Some hoped that scientific method would confirm at least the essence of these traditional deeply held beliefs, perhaps uncovering evidence that people lived who had special powers, channels to knowledge beyond the senses. Did the great religious teachers—Moses, Buddha, Jesus Christ, Muhammed—possess knowledge of a transcendent reality? Or were they like you and me, except more charismatic and thus able to attract bands of dedicated followers who then were able to convert great masses to their views by the powerful combination of word and sword? Can evidence for such special channels be found today, if we seek it systematically, carefully, and rationally?

It is not my intention here to examine evidence from the distant past. Gleaning any truth from data that have largely evaporated is too tough a job for me. I'm used to physics, with the data placed right before my eyes. So I prefer to concentrate on the present, or at least near-past, where the data are far more accessible. Surely, if supernatural forces exist they know no boundaries of time or space. If ancient people had revelations and other supernatural powers, people now walking the earth must also possess such powers, and a simpler task would seem to be to search out these exceptional individuals today rather than to seek to verify the truth of dusty ancient claims.

So let us look where most of the data lie in recent times, beginning back about a century and a half ago when some scientists turned their attention to the study of what appeared to be strong indications of the existence of supernatural phenomena.

Spiritualists and Mediums

In the middle of the nineteenth century, a wave of spiritualism swept America and Europe. Certain people came forward who seemed to possess remarkable powers. They appeared able to suspend natural law, and to have access to a special channel to the world beyond that had always proved so elusive. Suddenly this channel was not so elusive. Suddenly it seemed relatively accessible to thousands of people.

Furthermore, the phenomena demonstrated by these supposed superhumans involved real physical events that seemed to lend themselves to the same type of careful scientific study that had proved so stunningly successful for other aethereal phenomena: light, electricity, and magnetism. Perhaps the new phenomena were not supernatural. Perhaps they signaled the existence of natural forces just beyond the realm of normal experience that would make themselves more evident when examined carefully under the microscope of modern science. Certainly it was worth a look.

Spirit Rapping

The story starts in 1848, with two young girls in Hydesville, New York. Twelve-year-old Kate Fox and her fifteen-year-old sister Margaret became instant celebrities after claiming they could communicate with the spirit of a peddler who had been murdered in their house years before. These communications took place by means of coded rapping sounds that could be clearly heard by witnesses when the girls were present.

Dubbing themselves "mediums," they moved to Rochester where they performed before astonished audiences. Investigating committees composed of prominent citizens could find no explanation for the rappings, although a Rochester physician, Dr. E. P. Langworthy, and a handful of others, publicly stated their opinion that the noises were produced by the girl's feet (Kurtz 1986).

The debunkers did little to dampen the enthusiasm with which the Fox sisters were accepted by the public and the media. In 1850, the girls were exhibited by P. T. Barnum at his American Museum in New York along with Tom Thumb and the Bearded Lady.

As the Fox sisters gained fame and fortune, other individuals soon discovered, to their pleasant surprise and the benefit of their purses, that they too had spiritual abilities. Soon mediums were conducting seances in every major city in America, displaying powers that included not only spirit rappings, but actual vocal and visual manifestations from the world beyond. By 1853, only five years after the Fox girls' original discovery, some forty thousand spiritualists were located in New York City alone (Cerullo 1982). Thousands more were reported in other American cities. In 1852, the American mediumship craze spread to England, where it was quickly duplicated.

A Confessed Hoax

Years later, the Fox sisters confessed that they had perpetrated a hoax right from the beginning. They admitted that the rapping sounds were made, as Dr. Langworthy had suggested over thirty years earlier, by cracking their toejoints and other tricks as well as the use of accomplices. (For a reprint of Margaret Fox's confession, which originally appeared in the *New York World,* October 21, 1888, see *A Skeptic's Handbook of Parapsychology,* Kurtz 1985, p. 225).

Margaret Fox's confession did little to discourage spiritualism, which by that time had become a thriving business, not just for mediums but for anyone who could learn a few magician's tricks, gather an audience, and convince it that he or she had access to unseen planes of reality. As we continually find, this is an amazingly simple enterprise.

Secular Supernaturalism

The mediums and seances of nineteenth-century spiritualism offered both the average citizen and the educated elite a form of supernatural belief that was independent of traditional religion. This had great appeal to those who saw in spiritualism a way to confirm notions of immortality and the supernatural that did not rely on the now discredited credentials of traditional sacred authority.

Most scientists took an open-minded view, recognizing that realities may exist about which we have not yet gained knowledge except perhaps in glimpses by ancient and modern mystics. These scientists would be expected to look to science as the best way to investigate the possibilities.

So for all these reasons and others, a new, secular spiritualism, made respectable by serious scientific attention, spread like wildfire. After several transformations, it continues today in its latest incarnations: ESP, psychic powers, parapsychology, and the New Age.

Scientific Spiritualism

Although modern secular spiritualism started with the Fox sisters' prank in upstate New York, the scientific study of the phenomenon originated in England. There, a number of distinguished scientists hoped they might find in spiritualism a way to reconcile the conflict between their Protestant upbringing and science.

Mediums seemed to have the ability to communicate with the dead. If this could be verified, then the existence of the soul and life after death would become scientifically demonstrated realities. Despite the evolution of the human body from inanimate matter, the mind might still be shown to possess the immaterial, spiritual quality—the duality—that religious and philosophical traditions had taught. Surely such a scientific discovery would be the greatest of all time. Thus began what became known as psychical research.

Turning the Tables

One of the giants in the history of physics is Michael Faraday (1791–1867). His discovery of electromagnetic induction, in which a time-varying magnetic flux generates an electric current, was the major empirical step toward the unification of electricity and magnetism. Out of this came electromagnetic waves, as well as electric motors, generators, transformers, and much other modern technology.

Faraday was neither a fervent believer nor nonbeliever in spirits. Rather, he was an open-minded empiricist. In 1853 he applied his immense talents for developing scientific experiments to the problem of supernatural or psychic phenomena. One such phenomenon that seemed particularly amenable to scientific study was "table turning." People would sit at a round table, spread their hands out on the table, close their eyes, and concentrate, and without anyone consciously attempting to move it, the table would rotate one way or the other.

A similar effect is observed with the familiar ouija board, in which a triangular piece of wood, the planchette, can automatically trace out messages when touched by the fingertips. This effect is called automatic writing. Table turning and automatic writing can be done in one's own home, with trustworthy family members, so skeptics cannot always explain it away as sleight-of-hand trickery on the part of a skilled medium.

In his experiment, Faraday attached several layers of thick paper to a table top, using a soft cement that allowed some movement from one layer to the next. He then measured the angle of rotation of each layer when the table rotated during a seance. Here's what he found: the top layer, in contact with the sitters' hands, rotated more than the layer cemented to the table top.

Next, Faraday attached levers to the table, controlling a pointer that indicated whether the sitters' hands or the table moved first. When the sitters could see the pointer, no motion occurred. Thus, Faraday was able to show that involuntary muscular movements of the seance participants caused the rotation; when they had feedback on the motion of their hands, so that they could avoid these movements, nothing happened (Zusne and Jones 1982; Hyman 1989, p. 83).

Skeptics and Believers

Faraday's observations placed him firmly in the camp of the majority of British scientists, who were highly skeptical of the spiritualist claims, attributing them to natural effects, as with table turning, or trickery. As the fame of mediums spread, many were investigated and caught cheating. Professional stage magicians showed that they could duplicate most of the phenomena. Indeed many mediums, like modern psychics such as Uri Geller, had started their careers as professional conjurers. So while phony mediums and psychics can convince a gullible person that they have special powers, experienced magicians usually just chuckle.

Professional stage magicians normally do not claim supernatural powers. In fact, professional magicians, such as Houdini in those days and James Randi and Henry Gordon today, are at the forefront in exposing those who use magic tricks to try to fool people into thinking they possess paranormal powers.

In his 1924 book *A Magician Among the Spirits,* Houdini documented the fraudulent nature of mediums and told how he was able to perform most of their tricks naturally (Houdini 1924). Randi and Gordon have done the same in recent years (Randi 1986, Gordon 1987). To those who think irrational beliefs are harmless, take a look at Houdini's list of murders and suicides he claims were the result of seances (Houdini 1924, pp. 180-190).

During the heyday of mediums, the most outspoken scientific skeptics included distinguished figures such as John Tyndall, Francis Galton, William Thomson (Lord Kelvin), and Thomas Huxley. Interestingly, Darwin's codiscoverer of natural selection, Alfred Russel Wallace, blindly endorsed spiritualism.

Wallace believed in the duality of mind and body, with the mind separate from the natural world. However, unlike other like-minded scientists, Wallace applied no critical analysis to spiritualism, accepting everything he witnessed in seances at face value. His belief was unquestioning, and so, decidedly unscientific. As a result, he attained little credibility. In 1876 one of his favorite mediums, Henry Slade, was convicted of willful deception for profit.

In Wallace we see an example of an effect that continues to the present day, in science as well as pseudoscience: investigators often tend to let their own strong emotional stake in a particular outcome overrule their better judgment, and blind them even to facts that hit them between the eyes.

Despite the frequent public exposure of fradulent mediums and the lack of support from the mainstream scientific community, a few scientists in this period continued to pursue the study of psychic phenomena as manifested in the performances of mediums. They and their spiritualist supporters used an argument we still hear today among parapsychologists, UFOlogists, and other proponents of the paranormal: just because some people are fakes, so many examples of unexplained mysterious phenomena exist that "something must be there."

A number of British physical scientists of considerable prestige were included in the small group that studied psychic phenomena in the latter half of the nineteenth century. These included the Nobel Prize–winning physicists John William Strutt (Lord Rayleigh), the discoverer of argon, and Joseph John Thomson, the discoverer of the electron. Both were sympathetic to the aim of seeking evidence for the existence of mind as separate from matter. But, unlike Wallace and others, Rayleigh and Thomson did not allow their sympathies to interfere with their scientific judgment. Neither performed extensive studies of his own, but they provided encouragement and suggestions for those who carried out the actual work (Oppenheim 1986).

The Quest of William Crookes

The prominent scientist who pursued his own psychic studies with perhaps the greatest fervor was Sir William Crookes (1832–1919). Crookes was largely a self-taught chemist with a limited formal education. But he nevertheless made several major discoveries in chemistry and physics by dint of skillful experimentation conducted in a laboratory in his own house. He discovered the chemical elements thallium and selenium. He extensively studied cathode rays and invented a vacuum tube (Crookes tube) that was eventually used by J. J. Thomson to show that these rays were actually electrons. As we have seen, the electron is still regarded as one of the fundamental particles of matter. So William Crookes's contributions to physical science were of the highest merit.

After the death of his brother in 1867, Crookes was drawn into spiritualism and began an attempt to prove the reality of psychic phenomena using scientific methods and apparatus. Although Crookes's scientific career spanned decades, his five years of psychic research gained him greater, but unfortunately less flattering, historical recognition.

During 1871 Crookes held a series of seances with Kate Fox in London, after which he concluded that her spirit rappings "were true objective occurrences and not produced by trickery or mechanical means" (Crookes 1874). The same year, Crookes began a series of tests of the most skillful medium of the day, Daniel Dunglas Home.

Home had achieved an international reputation for his remarkable abilities in table turning, automatic writing, spirit materialization, and levitation. In the presence of Napoleon III in Paris, Home had levitated a table while his feet were being held, and materialized a ghost hand to write "Napoleon" in the handwriting of the departed conqueror (Cerullo 1982, p. 29).

In his most famous feat back in England, Home reportedly floated out of the window of a building and then back in another—though all in the dark, with everyone ordered to remain seated at the seance table.

Throughout his career, Home was never caught cheating, although reasonable natural explanations for his feats have been given (Hall 1984). After Home married a rich woman and retired comfortably, he exposed the techniques of less skillful mediums, but never admitted that he himself ever resorted to such chicanery.

Crookes's experiments with Home were not as carefully controlled as the scientist claimed or believed. This became evident later from published notes (Podmore 1963, Hyman, 1989, p. 91). As consistently happens when mediums and psychics are the subject of allegedly careful laboratory tests, Home was able to dictate how the tests would be performed. This breaks a cardinal rule of experimental method: the experiment must always be under the control of the experimenter.

The usual plea in psychic experiments is that psychic powers do not occur unless the conditions are right, as defined by the medium or psychic. The presence of excessively tight controls, "impersonal" laboratory equipment, and skeptical observers with their unfriendly thoughts is blamed for keeping the spirits away.

When, during a seance, Home would demand "all hands off the table," Crookes would comply along with the other sitters. Crookes gullibly swallowed ploys such as this and allowed Home to call the shots. Unsurprisingly, the result of this kind of poorly controlled experimentation was a continually impressive demonstration of paranormal powers. In 1871 Crookes announced that he had uncovered "the existence of a new force, in some manner connected with the human organization, which for convenience may be called the *Psychic Force*" (Crookes 1871).

Crookes Gets Cooked

In 1874–1875, William Crookes conducted a series of sittings with an attractive young medium named Florence Cook. Cook was supposedly able to materialize the full body of a spirit, who would emerge from behind a curtain during her seances and parade around the room to the wonder of the participants. Cook was behind the curtain when all this happened and was never observed simultaneously with the spirit.

Crookes made numerous tests in his own laboratory, where he could bring his excellent equipment to bear on the problem. He took many photographs. One of these can be found in Alfred Douglas's credulous history, *Extra-Sensory Powers* (Douglas 1977). It shows "Katie King," a spirit Cook has supposedly invoked, in front of the curtain. Above this picture is one of Cook herself. Obviously the two women are the same. The prominent sharp nose and distinctive shape of the lips are identical!

Crookes used other technological means to attempt to determine whether

Cook was using trickery. For example, he attached a galvanometer to Cook that was designed to detect any movements on her part. This failed to produce any evidence that Cook used fraudulent techniques. Crookes proclaimed that "Katie King" was unequivocally real (Crookes 1874, as reprinted in Barrington 1972). Later it would be demonstrated how, using other parts of her body or a resistance coil, Cook could have maintained the galvanometer current.

In 1880, Cook was caught impersonating her materialized spirit when someone grabbed her and tore open the curtains, revealing the medium's chair empty (Oppenheim 1985 and references therein).

With his zealous promotion of the reality of psychic phenomena, Crookes found himself under attack from the skeptical segment of the scientific community. He deserved criticism for his credulity, but some of the attacks were unfair and even slanderous, not always focusing on the scientific questions. Crookes's relationship with Cook was questioned. It was suggested that, taken in by her charms, he had joined with her to dupe the public (Hall 1963). More likely, he was duped.

In an unsigned article, later identified as being written by W. B. Carpenter, a physiologist and registrar at London University, Crookes's discovery of thallium is scoffed at as being "purely technical." The article goes on, "We are advised, on the highest authority, that he is regarded among chemists as a specialist of specialists, being totally destitute of any knowledge of Chemical Philosophy, and utterly untrustworthy as to any inquiry which requires more than technical knowledge for its successful conduct" (Carpenter 1852).

Stung by these attacks and perhaps having some second thoughts of his own, Crookes backed off from his claims and for the next forty years of his life had little more to say on the subject of spiritualism, though he remained active in the psychic research societies. Clearly, he continued to believe that some kind of unseen force existed, and in 1879 proposed a fourth state of matter beyond the gaseous state, based on his observations of high vacua. Today we know that an ionized plasma is another state of matter, obtained at very high temperatures when electrons are stripped from the atoms of a gas. So Crookes was right about a fourth state, but it has nothing to do with any "psychic force."

Shortly after his wife's death in 1916, the eighty-four-year-old Crookes accepted as legitimate a fake photograph of her spirit, an obvious double exposure by an unscrupulous photographer (Oppenheim 1986). Deep down, his desire to believe blinded him to the chicanery of his psychic subjects. William Crookes was a great physical scientist. His tragedy was that his wish to believe in a spirit world prevented him from applying his talents to his psychical investigations.

Sir Oliver Lodge

Another well-known and highly competent nineteenth-century physical scientist, whose fervent desire to believe in spirits led him to pursue psychic research with an insufficient component of skepticism, was Oliver Lodge. Lodge had contributed importantly to the development of radio (then called "wireless telegraphy") and had conducted experiments on relative motion that formed part of the database that led to the eventual development of special relativity.

With his scientific interest in wireless communication, Lodge focused his psychic studies on thought transference, or "telepathy." In 1890, he studied the Boston medium Laura Piper, who William James thought might be the long sought "white crow" of the supernatural.

Lodge witnessed Piper provide information about the sitters at her seances, who steadfastly insisted that she had no way of knowing these details of their lives (Oppenheim 1986). With her permission, investigators from the Society for Psychical Research searched her luggage and mail, but found no evidence that Piper had any sources for her information except that obtained during her trances.

Piper's spirit control was one "Dr. Phinuit," a Frenchman who never spoke French, which is like saying a Frenchman who never eats bread, drinks wine, or makes love. Lodge recognized that Dr. Phinuit often fished for the answers to questions, a common technique one sees with professional psychics today, but the scientist was still convinced that thought transference had actually taken place.

Ultimately, Lodge also became convinced that telepathy occurred, not only between the living, but also between the living and the dead. He felt that the immortality of the soul had been demonstrated by the power of mediums. His son had been killed in Flanders in 1915, and Lodge wrote that Raymond was communicating with his family from the beyond (Lodge 1916).

William Crookes also had found comfort in spiritualism after the loss of a loved one. The proof of immortality, rather than the determination of the truth about spiritualism, seems to have been the primary goal of many psychic researchers.

Like Crookes, Lodge was ridiculed for his beliefs. It was proposed that his mental condition was "indistinguishable from idiocy" and also— a more kindly criticism—that he was ready to accept unsound evidence because of "the will to believe" (Tuckett 1912).

The Royal Society

The Royal Society of London for Improving Natural Knowledge was incorporated in 1662. Samuel Pepys and Isaac Newton were among its early presidents, and latter presidents have included individuals we have already mentioned such as Lord Rayleigh, William Crookes, and Joseph John Thomson.

The Royal Society has served as the prototype of the societies that continue to play a major role in the scientific process in many countries. These societies sponsor conferences, provide industry and government with a source of independent expert consultation, and most important, publish the carefully refereed and edited journals in which scientific knowledge is primarily disseminated to the scientific community and archived for posterity.

The Psychical Research Societies

In the latter part of the nineteenth century, the journals of the Royal Society still were the most prestigious place to publish scientific papers. But the great explosion of science at that time triggered a proliferation of more specialized societies and journals to handle the great volume of new literature. When psychical research began to develop, the researchers did not find wide support within the scientific establishment, as represented by the skeptical membership of the Royal Society. Papers on psychical research submitted to the Society were usually rejected, so psychical researchers formed an uneasy alliance with the nonscientific spiritualist community to organize, in 1882, the Society for Psychical Research (SPR).

The first president of the SPR was Henry Sidgwick, a highly regarded professor of moral philosophy at Cambridge University. Sidgwick was able to attract some of the cream of European and American intelligentsia to become members of SPR, including William James, Henri Bergson, Arthur Balfour, and Charles Lutwidge Dodgson (Lewis Carroll).

Three years after the formation of the SPR, the American Society for Psychical Research (ASPR) was established by William James and others. In 1889 the ASPR was incorporated as the American branch of the British SPR, but this connection was dissolved in 1907, and the American organization continued independently. Both the SPR and ASPR are still in existence today, and both publish reports of psychical investigations.

Exposing Theosophy

In its finest hour, the British SPR sponsored an investigation of Theosophy and its founder, Madame Helena Petrovna Blavatsky. Theosophy is a mixture of Eastern religions and occult ideas from many nations, collected in Blavatsky's *Isis Unveiled* and the six volumes of *The Secret Doctrine*. She claimed to have received the revelations reported in those texts from ancient "mahatmas," including Buddha, Zoroaster, and Christ. Blavatsky introduced the term "astral projection" still in vogue today as the means by which "astral planes" of knowledge are reached in the world of the occult.

Messages often came to Madame Blavatsky in written form, magically appearing on her desk in the Occult Room of the Theosophical Society headquarters in Madras, India. At a shrine in the adjacent room, other miracles would occur, testified to by the usual impeccable witnesses. Broken china would reassemble itself and letters from the mahatmas to individual petitioners would float down from the ceiling.

In 1884, the SPR provided funds for a young investigator, Richard Hodgson, to travel to India to track down the source of Madame Blavatsky's mysterious revelations. While he did not set out to debunk Theosophy, and Madame Blavatsky was on good terms with the SPR leadership, Hodgson honestly reported his findings in the third volume of the SPR *Proceedings* in 1885.

With the help of handwriting experts, Hodgson demonstrated that the mahatma letters were written by Madame Blavatsky herself. Ultimately, the miracles of Madras were exposed as frauds. The tricks used were revealed in letters from Madame Blavatsky to her accomplices in Madras, explaining how to accomplish miracles in her absence. Narrow accessways were found between the Occult Room and the shrine, and other construction details made it easy to see how the tricks were accomplished (Oppenheim 1985).

Of course, this did not mark the end of Theosophy, any more than the confessions of the Fox sisters marked the end of spiritualism. The story of Theosophy is simply one more example of how easy it is to form a new religion, even in the modern era. One wonders if magic tricks may have been used by some of the original leaders of the great religions to convince their followers that they had special powers handed down from God. From what we know about the founders of religions closer to our time—Madame Blavatsky, Joseph Smith, and Ellen G. White, among others—we cannot have much confidence that the miracles of the founders of the great religions were not tricks that appeared supernatural to the eyes of simple, superstitious people (Kurtz 1986).

Compromising the Societies

In numerous public statements, Sidgwick and other SPR leaders insisted that their organization was a scientific society, dedicated to the use of scientific methods to establish the truth or falsity of psychic phenomena. In practice, the application of this principle often led to conflicts within the organization, as true believers objected to the SPR's rejection of reported positive results that failed to meet even minimal critical standards.

Sidgwick himself, despite his academic credentials, failed to understand the necessity of a hard-nosed, skeptical attitude when dealing with extraordinary claims. In his first presidential address after the formation of the Society in 1882 he said, "I say it is a scandal that the dispute as to the reality of these phenomena should still be going on, that so many competent witnesses should have declared their belief in them, that so many others should be profoundly interested in having the question determined, and that the educated world, as a body, should still be simply in an attitude of incredulity." He would be shocked to his boots to know that the educated world, as a body, maintains this attitude of incredulity over a century later.

Spiritualists outside the scientific community had previously formed several societies of their own, and while promising high standards for their investigations, in practice rarely applied them. Basically, they looked to science to provide support for something they already were sure existed.

Further, few spiritualists had any real understanding of scientific method. Like the average person today, they saw no reason not to accept a paranormal explanation for an event when the data on the event were inadequate to provide a conclusive natural explanation. They did not understand, or chose not to accept, the scientific rule that the burden of proof lies with those who seek to demonstrate any new proposition, and that the mere absence of sufficient data to provide an explanation based on known principles cannot be used as an argument for overthrowing those principles. As we have seen in the cases of William Crookes and Oliver Lodge, even scientists can be blind to mundane explanations for their observations if they have a deep psychological need to believe in more esoteric explanations.

Conflicts between the scientists and spiritualists within SPR and ASPR inevitably led to the fatal compromising of the integrity of both societies and of the journals they produced. The result has been a century of enormous but futile effort in which thousands of scientifically useless facts have been accumulated and published.

Why useless? Because of their anecdotal nature; because of the unreliability of human witnesses in uncontrolled circumstances; because of the lack of sufficient data to rule out conventional explanations; because of all the reasons the scientific method rejects data that do not meet critical standards, and conventional scientific journals reject submitted papers that fail to meet these criteria.

Investigations of mediums continued into the twentieth century, and the SPR and ASPR on occasion exposed blatant cases of fraud even their own credulous memberships could not swallow. But their journals have never succeeded in achieving a high level of credibility in the eyes of the rest of the scientific community.

Today the psychical research journals continue as forums for believers to press their ideas, to respond to the attacks of skeptics, and to attack the skeptics in return. Nothing is wrong with that, as long as the editorial bias is admitted. The volumes occasionally contain some respectable studies, but most articles usually begin with the assumption that psychic phenomena are demonstrated realities. Since this is a belief and not an empirical fact, one might be justified in viewing the SPR and ASPR today as religious rather than scientific institutions.

References

Abell, George O., and Singer, Barry. 1981. *Science and the Paranormal.* New York: Scribner's.

Barrington, M. R., Goldney, K. M., Medhurst, R. G., eds. 1972. *Crookes and the Spirit World.* London: Souvenir.

Carpenter, W. B. (unsigned). 1871. "Spiritualism and Its Recent Converts." *Quarterly Review* 131:301-353.

Cerullo, John J. 1982. *The Secularization of the Soul.* Philadelphia, Pa.: Institute for the Study of Human Issues.

Crookes, Sir William. July 1871. *Quarterly Journal of Science.* Reprinted in Barrington, et al. 1972.

———. 1874. Letters to the Spiritualist Newspaper. Reprinted in Barrington, et al. 1972. P. 22.

———. 1874. *Researches in the Phenomenon of Spiritualism.* London: Burns and Oates. P. 88.

Douglas, Alfred. 1976. *Extra-Sensory Powers.* Woodstock, N.Y.: Overlook.

Frazier, Kendrick, ed. 1986. *Science Confronts the Paranormal.* Buffalo, N.Y.: Prometheus Books.

Gardner, Martin. 1981. *Science: Good, Bad and Bogus.* New York: Avon.

Gordon, Henry. 1987. *Extrasensory Deception.* Buffalo, N.Y.: Prometheus Books.

Hall, T. H. 1963. *The Spiritualists: The Story of Florence Cook and William Crookes.* New York: Helix.

———. 1986. *The Enigma of Daniel Home: Medium or Fraud?.* Buffalo, N.Y.: Prometheus Books.

Hansel, C. E. M. 1980. *ESP and Parapsychology: A Critical Reevaluation.* Buffalo, N.Y.: Prometheus Books.

Houdini, Harry. 1924. *A Magician Among the Spirits.* New York: Harper Bros.

Hyman, Ray. 1989. *The Elusive Quarry: A Scientific Appraisal of Psychical Research.* Buffalo, N.Y.: Prometheus Books.

Kurtz, Paul, ed. 1985. *A Skeptic's Handbook of Parapsychology.* Buffalo, N.Y.: Prometheus Books.

Kurtz, Paul. 1986. *The Transcendental Temptation: A Critique of Religion and the Paranormal.* Buffalo, N.Y.: Prometheus Books.

Lodge, Oliver. 1916. *Raymond, or Life and Death.* London: Methuen.

Mauskopf, Seymour H., and McVaugh, Michael R. 1980. *The Elusive Science: Origins of Experimental Psychical Research.* Baltimore, Md.: Johns Hopkins University Press.

Oppenheim, Janet. 1985. *The Other World. Spiritualism and Psychical Research in England, 1850–1914.* Cambridge: Cambridge University Press.

———. May 1986. "Physics and Psychic Research in Victorian and Edwardian England." *Physics Today.* P. 62.

Podmore, F. 1963. *Mediums of the 19th Century.* New Hyde Park, N.Y.: University Books.

Randi, James. 1986. *Flim-Flam!* Buffalo, N.Y.: Prometheus Books.

Rawcliffe, D. H. 1959. *Illusions and Delusions of the Supernatural and the Occult.* New York: Dover.

Tuckett, I. 1912. *Bedrock* 1:204.

Zusne, Leonard, and Jones, Warren H. 1982. *Anomalistic Psychology: A Study of Extraordinary Phenomena of Behavior and Experience.* Hillsdale, N.J.: Lawrence Erlbaum Associates.

8.

Psience

When reputable scientists correct flaws in an experiment that produced fantastic results, then fail to get those results when they repeat the test with flaws corrected, they withdraw their original claims. They do not defend them by arguing irrelevantly that the failed replication was successful in some other way, or by making intemperate attacks on whomever dares to criticize their competence.
Martin Gardner, *Science: Good, Bad and Bogus*

The Scottish Verdict

J. J. Thomson, the discoverer of the electron, joined the SPR in 1883, shortly after its creation, and served on its councils for more than thirty years. Like many scientists at that time, he thought there might be some validity to psychic phenomena. But after a half century, Thomson remained unconvinced. In his memoirs he gave telepathy "the Scottish verdict—not proven" (Thomson 1936, p. 158).

More recently, after another half century of psychic research, the National Research Council (NRC) has similarly concluded that "the best scientific evidence does not justify the conclusion that ESP—that is, gathering information about objects or thoughts without the intervention of known sensory mechanisms—exists" (Druckman 1987).

The Hyperspacial Nuclear Howitzer

The NRC is the operating arm of the National Academy of Sciences, the American equivalent of Britain's Royal Society. Fellowship in the Academy is awarded to the nation's most distinguished and prestigious scientists. Operating independently of government, universities, and industry, the Academy offers advice and undertakes investigations on scientific matters of national importance and public interest, and often with controversial overtones.

In 1984, the Academy was asked by the U.S. Army to examine a wide range of techniques reported to enhance human performance. If the Army could speed up its training of soldiers, it could save millions of dollars each year. The techniques proposed included sleep learning, accelerated learning, integrating hemispheric activity, biofeedback, altered mental states, neurolinguistic programming, and ESP.

In the case of ESP, popular books had appeared suggesting that psychic powers might be put to military use. For example, an enemy might read the minds of commanders, or plant ideas inside them. A battalion of "warrior monks" might be formed, who were able to leave their bodies at will, levitate, and walk through walls. Suggestions for "psychotronic" weapons included the "hyperspacial nuclear howitzer," and an "antimissile time warp" that deflected incoming ICBMs and sent them into the past to explode harmlessly among the dinosaurs.

While these claims are bizarre, to say the least, requests for funding on psychic research were being partially justified on defense grounds, with the familiar argument that the Russians were doing it (see, for example, Targ and Harary 1984).

The NRC formed a committee of fourteen psychologists, neuroscientists, and other scholars and experts to conduct the investigation. The members made an intensive study of the available literature, visited laboratories, talked to leading investigators in the various fields, and commissioned analytical studies of their own.

While some positive results were reported in the use of certain practices such as stress control, no evidence was found to support many claims that were widely promoted in self-improvement books, tapes, and high-priced motivational seminars.

Ironically, of the various techniques surveyed, significant empirical data existed only for those based on the assumption of paranormal powers. Thousands of published experiments offer evidence for psychic phenomena. But the issue is not the magnitude of the database, but rather what conclusions can be drawn from it. The Academy report concluded that "even the most solidly based arguments for the existence of paranormal phenomena fall short of currently accepted parapsychological standards." In other words, despite the huge data sample, ESP is not proven even according to criteria established by parapsychology.

The Paranormalists Fight Back

As you might expect, many promoters of psychic phenomena do not agree with this assessment (see, for example, Palmer 1989 and Alexander 1989). They complain that the committee was biased. But scientific method demands a bias in favor of the conventional explanations for a phenomenon until convincing evidence is provided for an unconventional one.

The critics of the report also complained that the committee failed to find convincing alternative nonparanormal explanations for many of the studies. But in doing so these critics ignore the scientific rule that the burden of proof lies with the proponents of a new phenomenon, not those who critically review the evidence.

The psi spokesmen disagreed with the weight placed by the NRC on the failure of psychic phenomena to replicate, arguing that replication is not a requirement in the behavioral sciences. Then so much the worse for the behavioral sciences! As I have already noted, paranormalists would like to rewrite the rules of science so that their results can be admitted. That will always be vigorously resisted by most scientists.

The bottom line is simple: science is based on consensus, and at present a scientific consensus that psychic phenomena exist is still not established.

J. B. Rhine and ESP

The term "Extrasensory Perception" was coined by the central figure in psychic research in the twentieth century, Joseph Banks Rhine. For many years, Rhine headed the Duke Parapsychology Laboratory that he had founded in 1940 at Duke University.

A botanist with a strong religious background, Rhine, like his nineteenth-century predecessors, saw psychic research as a possible bridge between science and religion. Upon getting his Ph.D. in botany from the University of Chicago in 1925, he decided that psychic research, which he renamed parapsychology, offered him more exciting opportunities.

While visiting Harvard in the summer of 1926, Rhine and his wife Louisa, also a Ph.D. botanist, attended a seance conducted by the famous Boston medium Margery Crandon. The Rhines' scientific sensibilities were properly shaken by the observation of what was, to them, rather obvious trickery (Mauskopf et al. 1980, p. 76).

In 1927 J. B. Rhine went to work in the psychology department at

Duke. Two years later, he wrote a paper on the supposedly telepathic powers of a horse named Lady Wonder, stating that "only the telepathic explanation . . . seems tenable in view of the results" (Rhine 1929). Later it was discovered that the owner had been using subtle signals to control the horse's behavior (Hines 1988, p. 84).

These disappointing experiences did not discourage Rhine from continuing his psychic research, but they convinced him to change the direction of the research away from the traditional investigations of anecdotal stories, mediums, and haunted houses.

As a natural scientist trained in the methods of the laboratory, Rhine realized that the laboratory provided an environment where more careful controls could be maintained over the many extraneous factors that often confuse observations. In no time at all Rhine had results from his laboratory experiments. His 1934 book *Extrasensory Perception* presented preliminary studies that he said gave overwhelming evidence for the existence of ESP (Rhine 1934).

Rhine distinguished between three forms of ESP: (1) *telepathy* or thought transference; (2) second-sight or *clairvoyance;* and (3) looking into the future or *precognition.*

While any number of variations of experimental techniques were tried at Duke over the years, Rhine's typical experiment involved simple card-guessing. Using the now-famous Zener, or ESP, cards that contain five presumably neutral geometrical figures (a square, circle, star, cross, and three vertical wavy lines), the subject tried to guess which card would be selected from a shuffled deck typically of twenty-five cards. The subject's choices were then compared with what would be expected by chance.

Rhine reported results that were highly unlikely to have happened by chance, by odds of billions to one. This remarkable success, unprecedented in psychic research history, made him instantly famous.

Critics quickly observed, however, that the experiments were far from airtight. Although challenges of Rhine's statistical techniques were unable to explain all the data, other aspects of his experiments remained questionable. In particular, numerous opportunities for sensory clues to his subjects were shown to exist. Further, the subjects had an incentive to find ways to beat the odds. They were generally paid student help, and the successful ones kept their jobs, on the presumption that they possessed ESP abilities that should be tested further. Sometimes the subjects were paid on the basis of the number of correct responses.

In the typical experimental setup, the subject was not adequately isolated; thus opportunities for cheating could not be ruled out. Furthermore, flaws in the design of some types of ESP cards made their symbols detectable from the reverse side. Sensory detection was also possible by other means familiar to the card shark. (For a complete critique of these experiments, and others that came later, see Hansel 1980.)

Rhine insisted that the results presented in *Extrasensory Perception* were only "early and minor," though they were hardly interpreted as minor by the media. In following years, he made serious attempts to develop tighter procedures. Even these procedures were often not tight enough, however, but when they were, only negative results were found.

Rather than accepting these negative results as evidence against ESP, and concluding that his preliminary results were invalid, Rhine attributed them to a "decline effect": ESP ability somehow fell off with time as subjects became bored.

Thus, Rhine betrayed the same weaknesses of the believer we saw in William Crookes and Oliver Lodge, and the same credulous, trusting nature. Despite his good intentions in providing a controlled laboratory environment, Rhine showed an unwillingness to apply Occam's razor. A much more economical hypothesis for explaining the decline effect is that the phenomenon never existed in the first place, the original positive results having been spurious, and due to inadequate controls. But Rhine was not the impartial investigator, searching out the truth wherever it might lead. He believed in ESP, and set out to prove it, just like the spiritualists of the nineteenth century.

The Pearce-Pratt Experiment

The Pearce-Pratt experiment, also known as the Campus Distance Series, which was conducted from August 1933 to March 1934, is a classic in the history of parapsychology. This particular experiment was partially motivated by a comment of Einstein, who had said that he had an open mind on ESP, but would not believe it until he saw a distance effect (for a review of Einstein's statements on ESP, see Gardner 1981).

The "mental energy" radiated in ESP should fall off with distance, as does light or sound energy. Radiation that is not focused, but spreads in all directions with equal intensity, is uniformly distributed over the surface of a sphere whose area increases as the square of the distance. Thus, the signal intensity, or energy per unit area, decreases as the square of the distance from the source. Even highly focused energy, such as that of a laser beam, will spread somewhat and become less intense at greater distances.

Rhine instructed his associate J. G. Pratt to conduct experiments with one of the star performers at the Duke laboratory, divinity student Hubert Pearce. Pearce had performed well in the preliminary tests, especially when no one else was in the room. His scores tended to fall off when someone dropped by to watch, a common occurrence that parapsychologists attribute to distraction or other changes in the laboratory atmosphere. This is known as the "observer effect." Interestingly, the observer effect is greatest when the observer is a skeptic or professional magician.

In a typical run of the Campus Distance Series, Pearce synchronized his watch with Pratt's and then was supposed to go to either of two locations, 100 and 200 yards from Pratt's office. At an agreed time, the experiment would begin.

Without looking at the cards in two shuffled decks of twenty-five cards each, Pratt placed one card face down on the desk in front of him each minute. As each card was placed on the desk, Pearce was to write down his guesses as to the symbol on the card. After fifty cards, Pratt turned the cards over and recorded the sequence. Rhine did not participate, except to collect the data sheets from the two participants and place them in a safe for later analysis.

In a total of seventy-five runs, three times Pearce achieved scores as high as thirteen of the twenty-five cards in a single deck. Since five different symbols were used, only five correct responses would be expected on the average from chance. The chance probability for thirteen or more correct guesses for twenty-five cards and five symbols is less than one in ten thousand. Other scores were lower, but the odds against Pearce's overall performance for the seventy-five runs being achieved by chance were 10^{22} to one.

Although several reports on the Pearce-Pratt experiment were published, it was not until twenty years later that a fairly full account of the experiment became available (Rhine 1954). In the meantime, the Pearce-Pratt experiment was widely touted as one of the best examples of evidence for ESP. But although the results were remarkable, they did not show the hoped-for distance effect.

Far from being discouraged by this failure to meet the Einstein criterion, ESP enthusiasts interpreted the absence of any dependence on distance as evidence for the supernatural quality of ESP: it was not bound by the laws of physics. Rhine himself always made it very clear that he believed ESP was a spiritual, that is, nonmaterial, phenomenon.

After the successful series of tests was completed, Hubert Pearce suddenly lost his powers. He was taken to greater distances and different places, but only negative results were obtained. Was this the distance effect? Apparently not, because Pearce never again demonstrated any significant ESP ability at short distances or long. For Rhine, another example of the decline effect was found. A more economical explanation is suggested by the fact that the Duke researchers gradually tightened their controls against trickery.

However, if the lack of a distance effect is an indication of the supernatural qualities of ESP, showing ESP's independence of the physical quantity of space, then the presence of the decline effect is an argument against supernatural qualities, since they should also be independent of the physical quantity of time.

The ESP Craze

With the astounding successes of Rhine and his collaborators at Duke, ESP became a craze rivaling the spiritualism of almost a century earlier. Spurred on by John Campbell, the publisher of the pulp magazine *Astounding Science Fiction,* few science fiction writers in the following decades failed to include mind-to-mind communication in their fantastic visions of the future. In more recent times, the movie and TV hits *Star Wars* and "Star Trek" have relied heavily on plot lines involving ESP. With this impetus from the mass media, most people today still assume that telepathy is a scientifically verified phenomenon.

Attempts to Repeat

With the attention given to the Duke lab, others jumped on the bandwagon and repeated the Rhine experiments. In the period 1934 to 1940, at least thirty-six experimental reports appeared in various journals. Some positive results were reported, but most came up empty. For example, six independent researchers with 500 subjects in a half-million trials obtained nothing but chance effects (Zusne 1982).

In a 1939 review of experimental results from 1934 to 1938, J. L. Kennedy concluded that only three positive experiments could not be explained by insufficient controls such as sensory clues and recording errors. These were the Pearce-Pratt and Pratt-Woodruff experiments at Duke, and an experiment of S. G. Soal in England (Kennedy 1939). Much later, Rhine himself concurred, saying that these were among the best three or four experiments prior to 1965.

Often skeptics are criticized for setting up straw men, criticizing weak claims that even parapsychologists disown. I agree it is only fair and proper to focus on those experiments that the experts in a field single out as definitive. This is the case for the experiments discussed in this chapter. All have been held by the parapsychological community, at one time or another, to be unassailable paragons of their field as primary evidence for the psi phenomenon.

Throughout the forties and fifties, it was widely believed, even in the scientific community, that evidence for ESP had been found. Although the three experiments mentioned were about the only ones regarded as "solid," so many other marginal positive effects were reported that parapsychology seemed primed for acceptance as a legitimate science. Surely the next generation of experiments, with improved techniques, would open up this marvelous new world beyond the senses.

Feeling that the existence of psi was pretty much established, parapsychologists directed their experiments toward studying the detailed properties of the phenomenon. Rhine and his colleagues in the U.S. and abroad began experiments to distinguish between the three forms of ESP (telepathy, clairvoyance, and precognition) and a fourth purported power of the mind, psychokinesis (PK): the ability to move objects mentally, that is, the power of mind over matter. Despite the enthusiasm, effects reported by experimenters were rarely confirmed by others and many skeptics remained unconvinced.

Hansel Shoots Down Pearce-Pratt

In 1960, C. E. M. Hansel of Manchester, England, visited Duke to take a hard look at the laboratory's experimental procedures, including those used for the Pearce-Pratt experiment, which had been performed over a quarter century earlier. As mentioned, the details of Pearce-Pratt did not appear in print until twenty years after the experiment was conducted, during which time Rhine and company had repeatedly asked the world to take their word that the experiment had been foolproof.

However, when Hansel examined the details of the Pearce-Pratt experiment and surveyed the site where it was conducted, he discovered a number of glaring deficiencies. These he expounded in an excellent book, *ESP: A Scientific Evaluation* (Hansel 1966; updated in *ESP and Parapsychology: A Critical Reevaluation,* Hansel 1980. For a later edition see *The Search for Psychic Power: ESP and Parapsychology Revisited,* Hansel 1989.)

Hansel points out that the subject, Hubert Pearce, had not been watched during the experiment. During the series of runs of the experiment, no checks were apparently made on his movements; no independent evidence was ever presented that he actually made all his guesses at the times and places specified in the experimental protocol.

Further, Hansel discovered that Pratt's desk could have been visible from outside the room in a number of plausible ways: from the corridor through the clear glass transom above the door, through the transom of the room across the hall, through the window in back of the desk, and through a trap door in the attic above that could have had a peep hole. Replicating the experiment, Hansel was able to achieve twenty-two correct responses for twenty-five cards by looking through the crack at the top of a door in an adjacent room, while an unsuspecting Duke staffer went though the identical procedure Pratt had used, overturning the cards and recording their sequence.

Hansel's critique, as you might expect, was not met with much enthusiasm in the parapsychological community. Here, after some thirty years, one of the pillar experiments of their field had been shown to be made of sawdust.

They fought back, arguing that Hansel hadn't provided any "proof" that Pearce had cheated, and complained that the figure in Hansel's book showing the plan of the rooms in the Duke building was labeled "not to scale." Did that matter or not? Hansel had tried to get the architect's plans from Duke, but had been rebuffed. If Hansel's scale was so far incorrect as to negate his argument, why not produce the plans to demonstrate it? Rhine's lab never demonstrated that Pearce could not have cheated in the ways proposed by Hansel.

Hansel did not have to prove anything. The burden of proving that cheating was impossible rested with Rhine and Pratt, not Hansel. Hansel succeeded brilliantly in exposing the. shoddiness of the experimental procedures of Rhine's laboratory.

Any number of simple precautions could have been taken to guard against fraud. In any sensible experimental protocol, Pearce would have been watched. Pratt's room should have been carefully sealed. Many other obvious precautions with the handling of the data were not taken, giving Pearce or Pratt a number of opportunities to change the figures.

Hansel Shoots Down Pratt-Woodruff

The Pratt-Woodruff experiment, carried out in 1938-1939, was highly touted by J. B. Rhine and others as the one with greater safeguards against error and fraud than any parapsychological experiment so far (Pratt and Woodruff 1939, Rhine 1954). Perhaps, but they were still far from adequate.

In this case, a screen separated the participants, which supposedly guarded against visual clues. Now as any bridge player knows, numerous ways exist for partners to signal one another, visually or otherwise. A screen removes the opportunity for providing visual clues, but does nothing to prevent other means by which sensory data can be transmitted. Sight is only one of five senses. Only complete isolation can eliminate all sources of natural data transfer.

The participants in the Pratt-Woodruff experiment still sat unnecessarily close together. A screen does not block sound. If you think about it a bit, you can easily come up with many ways to use sound to clue a confederate: throat clearing, changes in breathing rate, apparently innocuous verbal statements, scraping a foot on the floor, to mention a few.

Hansel duplicated the Pratt-Woodruff experimental procedure in his own laboratory in England and demonstrated how even an untrained person could obtain sufficient information from noises and shadows to produce odds inconsistent with chance.

For a fuller discussion of all the possibilities for clueing in the Pratt-Woodruff experiment, see Rawcliff 1959 and Hansel 1980. These challenges to Pratt-Woodruff have not been effectively countered, and so by themselves serve to rule out the results as evidence for ESP. However, there is much more.

The Pratt-Woodruff experiment tested a total of thirty-two subjects. On the subject's side of the screen, key cards containing the five ESP symbols were hung from pegs. The key cards were invisible to the experimenter on the other side, who held a shuffled deck of ESP cards. Blank cards visible to both sides were placed directly below the five pegs in a two-inch-high slit at the bottom of the screen. Note that the presence of this slit meant that total visual isolation was not accomplished.

In the experiment, the subject would signal his or her guess of the top card by pointing with a pencil through the slit to the blank card that was below the corresponding key card. Note that this provided the possibility for a visual clue, such as by the orientation of the pencil. The experimenter then placed the top card, still face down, directly next to the blank card that had been indicated. After twenty-five trials, the experimenter recorded the cards in each pile, while an independent observer on the subject's side reordered the key cards on the five pegs.

One subject scored 947 hits in 4,050 trials, compared to 810 expected by chance. The odds against obtaining 137 hits above chance in this many trials are more than twenty million to one. In 60,000 trials, the thirty-two subjects obtained at total of 12,489 hits, while 12,000 would be expected by chance. The chance odds against getting 489 or more excess hits are more than a million to one. Something beyond chance was undoubtedly operating—but was it ESP?

Most ESP experiments of the type pioneered by Rhine include numerous trials, numbering in the thousands. This would seem to provide assurance for the results. In many other branches of science experiments are often limited by the number of trials. However, a high number of trials introduces another problem that is illustrated in this example.

Note that the subjects would need to cheat only once in every 123 trials (60,000/489), less than one percent of the time, to achieve the reported significant inprovement over chance results. In experiments with large numbers of trials, the greater the number, the tighter the safeguards needed to rule out cheating and other sources of error.

In examining more carefully the data from twenty-two runs, Hansel discovered an interesting fact. The high-scoring cards had occupied the end key card positions of the previous run in seventeen cases, rather than the expected eight or nine. This suggested the following scenario: The subject would learn the key card sequence when the screen was taken down to record the sequence after a run. He or she could then figure that the card at one end or the other would be the first one replaced in the new sequence, since the cards would likely be removed in order. Remember this trick did not have to work each time—only one percent of the time to make a dramatic difference in the statistical significance.

Taking the 4,025 trials of the highest-scoring subject, Hansel showed that far more hits occurred—at odds of 100 billion-to-one against chance—when the card had previously occupied an end position. Hits on cards that had previously occupied one of the middle three positions occurred with only two-to-one odds, consistent with chance (Hansel 1980).

Three Decades of Nothing

So, through the persistent efforts of C. E. M. Hansel, the two experiments conducted by the Duke Parapsychology Laboratory, which for three decades were regarded as the best evidence for ESP by the parapsychology community, were shown to be fatally flawed. This rules them out as providing any evidence for ESP.

Rhine and others had continually claimed during that period that these experiments were beyond reproach, because of the careful techniques and complete safeguards against the possibility of fraud.

What if Hansel had not taken the trouble to read the Rhine papers, visit the Duke laboratory, analyze the raw data more carefully than the experimenters themselves, and set up his own tests? The Duke results would still exist in the literature, viewed as anomalies and touted by believers as evidence for ESP.

In over thirty years, the net product of the Duke Parapsychology Laboratory provided no believable evidence for ESP, despite tremendous publicity and public attention. The experiments showing the greatest positive effects simply were not sufficiently safeguarded against cheating. This is not to say that cheating was proven in the examples cited. No evidence has ever come to light that J. B. Rhine personally tampered with data or committed scientific fraud.

However, Rhine's laboratory was not totally free of the blemish of detected fraud. In the 1970s, after Duke University had severed its connection with Rhine's laboratory, which then reorganized as a private institute, Rhine's chosen successor as director of research, Walter J. Levy, Jr., was caught fabricating data by other staff members and duly fired.

Soal Gets Snagged

The third of the mid-century experiments cited as providing conclusive evidence for ESP was the Soal-Goldney experiment in England. This case has many fascinating aspects that have been examined in some detail by Kurtz (1986).

Samuel George Soal had been doing psychic experiments about as long as Rhine, trying to replicate the famous American's results with card guessing. In the period 1934–1939, Soal made 128,350 trials with 160 subjects, including a number of mediums, psychics, and spiritualists, but obtaining only negative results (Soal and Bateman 1954).

After this huge but disappointing effort, Soal was ready to give up on ESP. He wrote, "I have delivered a stunning blow to Dr. Rhine's work by my repetition of his experiments in England. . . . There is *no evidence* that individuals guessing cards can beat the laws of chance" (as quoted in Thouless 1974).

However, in 1939 Soal was convinced to go back and reanalyze his data. When he did so, he discovered that two of his subjects, Basil Shackleton and Gloria Stewart, scored significantly above chance expectations when their selections were compared with the cards just *before* or just *after* the target card. The target selections originally analyzed were consistent with chance. This "displacement effect" could be attributed to precognition or, I suppose, postcognition.

In 1941–1943, during the London blitz of World War II, Soal and Mrs. K. M. Goldney carried out a series of forty tests with Shackleton (Soal and Goldney 1943). Further tests with Mrs. Stewart were also conducted, in 1945.

In the 1941–1943 series, the subject and experimenters were in separate rooms. Although the door between the two rooms was left ajar, a questionable procedure, the subject could not see the person looking at the cards. I won't attempt to summarize all the results. Suffice it to say that, in the precognitive mode, where Shackleton's selection was compared with the next one selected by the experimenter-agent in the other room, he scored 1,101 hits in 3,789 trials. The chance odds for this result are an incredible 100 billion to one. Mrs. Stewart's results were even more impressive, as great as 10^{70} to one over chance.

The Soal-Goldney experiments were widely hailed and Soal was named president of the Society for Psychical Research in 1950.

In 1956, C. E. M. Hansel carefully critiqued the Soal-Goldney experiments and found a number of possible explanations (Hansel 1980). But the most interesting criticism was produced by George R. Price in an article in the American journal *Science*. The author applied philosopher David Hume's argument concerning miracles. Price agreed with Hansel that a number of ways existed for Soal to have cheated with the help of collaborators; therefore, this was a more likely explanation of the results than the "miracle" of ESP (Price 1955).

Soal, Rhine, and many others strongly protested Price's accusation, saying he had not proven that Soal had cheated. This was true. Price simply argued that the opportunity for fraud had not been adequately ruled out by the experimental procedure, making it the more probable explanation for Soal's results. Despite the protestations of unfair accusations against a respected scientist and gentleman, and even a later apology by Price, the fraud hypothesis was soon confirmed. Soal had almost certainly altered the data, but apparently did it alone rather than with the help of collaborators. Hansel and Price were wrong about the particular method used, but correct in their basic intuition that something fishy had occurred.

During the course of the Shackleton experiment, one of the experimenter-agents reported to Goldney that she had observed Soal changing 1s to 4s and 5s on the data sheets. Goldney told Soal, who promptly fired the agent. The charges remained unsubstantiated until 1971, when R. G. Medhurst showed that an excess of hits on 4 and 5 existed in the data from one of the two sittings with the fired agent (Medhurst 1971). Medhurst's accusations were confirmed in 1974 by a further study (Scott 1974).

Soal continued to deny the accusations, and prominent parapsychologists rushed to his defense once again. As before, the main argument was: "Where's the proof?" Soal had unfortunately lost the original data sheets that could have been examined for evidence of tampering. Can you imagine Carlo Rubbia losing the raw data tapes for the work that won him the Nobel Prize?

Convincing proof that data tampering had taken place in the Soal-Goldney experiment was obtained by Betty Markwick in 1978. Markwick carried out a sophisticated computer analysis to locate the pseudo-random number sequences that Soal had used. She discovered that extra digits were occasionally inserted in the sequence, and that these cases corresponded to hits. When the suspicious digits were removed, the hit rates fell to values consistent with chance (Markwick 1978).

The Dirty Test Tube

Although it has taken decades, the experiments of Rhine and Soal and their collaborators, regarded by many as providing conclusive evidence for ESP, were ultimately discredited. Today no experimental evidence for ESP remains untainted.

Nevertheless, for the sake of argument, suppose that one day a convincing set of replicated experiments supporting ESP is obtained. Then people will be tempted to go back and argue that Rhine and other early investigators had glimpsed something.

However, I must add a note of caution. The early experiments were so flawed in design and execution that they cannot be used to infer anything substantial about psi phenomena. These data should be excluded, just as the contents of a dirty test tube should not be added to those of a clean one before performing chemical analyses.

Yet these invalid data are used by many parapsychological authors and researchers. Although ESP remains unverified, they refer to the "decline effect" and "observer effect" and other reported properties of ESP as if these effects were real. Psi experiments are often designed to maximize these properties. For example, hired subjects are often released when their scores drop off. And experimental protocols frequently allow subjects to operate without supervision or to control many aspects of the experiment. The assumption is that psi only operates when people are "in the mood" and in an environment free from the bad "vibrations" emitted by skeptical observers. But, as they say about computers: "Garbage in, garbage out."

Automated Experiments

Using computers, which became available in the 1960s, many of the operational difficulties with the Rhine era experiments can be overcome. Errors in data recording and prospects for experimenter cheating can be minimized—though as we will see, not totally eliminated—by computers and other electronic instrumentation. The random number sequences used to generate target selections can be vastly improved upon by electronic generation. And with electronic communication, subjects can be more effectively isolated from the possibilities of sensory leakage.

Actually, technology that already existed in the Rhine era, such as the telephone and electronic counters, could have been used to improve the Duke experiments. But for reasons of personal taste and perhaps a basic mistrust of what Rhine called "physicalism," his psi experiments were conducted in only the crudest low-tech fashion, with data recorded laboriously by hand, and with random numbers taken from printed lists.

One particularly well-conducted electronic experiment was done at the U.S. Air Force Laboratory in 1963 (Smith et al. 1963, Hansel 1980). An apparatus called VERITAC automatically generated random target digits from 0 to 9, registered the subjects' guesses, and analyzed the scores. Tests of clairvoyance, precognition, and general ESP were conducted. The subject sat in a room with closed doors, well isolated from the VERITAC control console and peripheral equipment. A total of 55,500 trials were conducted with thirty-seven subjects. None exhibited any ESP ability.

Tart's Tarnished Trainer

The use of machines can provide pitfalls as well as advantages. One notable example is provided by an experiment of Charles Tart of the University of California at Davis, reported in his book *Learning to Use Extrasensory Perception* (1976). Using a device called the "Ten Choice Trainer," Tart claimed an ESP research breakthrough.

In order to minimize the possibilities of sensory leakage, subject "senders" and "receivers" were located in isolated booths. The sender in one booth faced a circle of ten playing cards with a light beside each. The machine randomly selected a card and turned on the adjacent light. The sender concentrated on that card and then pressed a button to signal the receiver in the other booth to make a choice. The sender also was provided with a TV picture of the receiver's console, which contained the same circle of cards and a button next to each. Thus the sender could try to "will" the correct choice.

Tart reported results that were far from chance. Skeptics pointed out several problems, however, including the use of signaling by time delay variations (Randi 1986). Recall that such a signaling system need not be perfect. Only a small percentage of hits achieved by means other than guesswork are sufficient to provide results with astronomical odds and a simulated positive ESP effect when the number of trials is very high. High numbers of trials become even easier to achieve with the use of automated equipment, so the controls must be tighter accordingly.

Tart's experiment had even more fundamental defects. Sherman Stein, a mathematician at the University of California at Los Angeles, analyzed the raw data and noticed something strange in the target selection statistics. Of 5,000 digits produced by the machine, only 193 pairs of repeated digits were found.

A common belief exists among gamblers that, in games of chance, the outcome of one turn is unlikely to be the same as the previous one. For example, with the roulette wheel, few gamblers will put money on the previous number, though in fact that number is as likely to occur again as any other. This "gambler's bias" is one of the statistical edges that help casinos maintain a steady, reliable income.

When random numbers from 0 to 9 are being selected, and 3 is chosen, the chance of the next number being 3 is still one in ten, the same as any other digit. In the case of the 5,000 trial digits in Tart's experiment, 500 pairs of the same digit should have occurred. Instead only 193 were found. Some flaw in the experimental arrangement resulted in over 300 fewer pairs, totally invalidating Tart's calculation of the significance of his results based on chance.

We can easily see how this experimental bias could have happened.

After the receiver made a choice, the sender would press a button to signal the computer to make the next selection. If the same card was chosen, which should happen one in ten times, the light adjacent to that card would remain on (the obvious precaution of shutting off the light and then turning it back on was apparently not taken). As a result, the sender might occasionally think that he had not pressed the button hard enough, and might press again for another selection. Or, seeing the same card light up, he might press again because of the gambler's bias mentioned above.

In order to answer some of the objections raised by skeptics, Tart made several improvements on his original device for the next series of experiments. Although he claimed in an article in *Psychic* magazine that "a big step toward repeatability of ESP" had been taken (Tart 1976), the results of the improved experiments were consistent with chance (Tart 1979).

Rather than admit that the original results published in his book were wrong, Tart blaims his University of California at Davis student subjects for "a dramatic change in attitudes." He came up with a remarkable explanation: "In the last year or two, students have become more serious, competitive, and achievement-oriented than they were at the beginning of the first experiment. Such 'uptight' students are less compatible with strong interest and motivation to explore or develop a 'useless' talent such as ESP" (quoted in Gardner 1981).

As with the decline and observer effects, we once again see a parapsychologist taking a null result and reinterpreting it as a property of a phenomenon that has not yet been shown to exist. I suppose succeeding experimenters are now careful to pick more idealistic students as subjects, now that Tart has demonstrated that they have a greater tendency toward psychic abilities.

Pseudo-Random Numbers

Before the advent of computers, random numbers were chosen from tables. Today we can generate them with computers. However, in both cases, the numbers are not truly random, but based on some mathematical algorithm that ultimately gives a repeatable sequence to the numbers produced. The numbers are "pseudo-random."

Although the actual sequence is complicated, if pseudo-random numbers are used in the experiment, it is possible that a subject can learn the algorithm used to generate them. For example, if the test is being conducted on an IBM-compatible personal computer using the MS-DOS random number generator, and the subject is able to invade the testing program to discover the algorithm used, he or she can simply duplicate the sequence on any available machine.

Simple precautions can help guard against this prospect. For example, the random number sequence begins at a place determined by an initial input number called the "seed." Seed selection can be done carefully, in secret, by several disinterested persons, so neither the subjects nor the experimenters have access to this number at any time prior to the experimental run or during the analysis of the data.

Quantum Processes Predicted?

Another approach was taken in the 1960s by Helmut Schmidt, then employed by Boeing Research Laboratories. Schmidt developed a random number source that was based on the fundamental unpredictability of the times at which radioactive nuclei decay. Schmidt put a Geiger-Mueller tube near a Strontium-90 radioactive source and used its output to start a circuit that counted the pulses of a free-running oscillator, selecting four possible outputs.

The conventional interpretation of quantum mechanics holds that processes such as atomic and nuclear transitions happen randomly, that is, in an undeterminable, unpredictable fashion. So radioactive decay should be a generator of purely random numbers. Even the most skilled charlatan cannot obtain a priori knowledge of the target sequence, since such information does not exist.

Schmidt went further, however. He argued that if a person succeeded in achieving a significantly better-than-random success rate using his device as the target selector, then a major violation of a fundamental physics principle would be demonstrated. The human mind would have achieved miraculous control over a physical process.

In an article in *New Scientist* in 1969 entitled "Quantum Processes Predicted?" Schmidt claims precisely this (Schmidt 1969). Subjects were asked to predict which of four lights selected by the random number generator would flash. Electronic counters for the numbers of hits and attempts were built into the apparatus, though they could be switched off and did not record every trial (Hansel 1980).

One subject, Rev. Keith Milton Rhinehardt, achieved a scoring rate whose chance probability was less than one in 500 million. Schmidt insisted that the tests were absolutely fraud-proof. He claimed his results demonstrated a "correlation between the future state of the random target generator and the present state of the subject's mind" (Schmidt 1969).

Once again, the most detailed critique of Schmidt's work can be found in Hansel's book (Hansel 1980). He lists a large number of deficiencies in the data-collection procedure that made it impossible to obtain a complete record of all the trials and their results. Insufficient raw data were provided for independent analysis. The result was that reliable calculations of the chance probabilities were impossible.

But perhaps the greatest flaw of the Schmidt experiment—a fatal one—was that it was performed by a lone experimenter. No one was watching Schmidt. Now I hasten to add I do not accuse Schmidt of cheating. Rather, I simply point out a defect in his experimental procedure. With the history of psychic research marred as it is by so many proven instances of fraud on the part of subjects or experimenters, no psi experiment conducted by a lone experimenter will ever be credible to the majority of the scientific community. This may strike the reader as unfair, but it is a fact.

Now, no experiment can ever be made totally immune from fraud. It's a question of degree. For the results of an experiment to be acceptable to the scientific community, fraud must be a less economical explanation than the explanation that some new phenomenon has taken place. Before we can allow the consideration of new hypotheses, the fraud hypothesis must be made exceedingly unlikely. And even when this criterion is met, we still require independent replication.

In Schmidt's case, he could have taken any number of precautions to ensure that he was reasonably free of suspicion of fraud. Other scientists, or preferably magicians, could have been asked to oversee the experiment. Schmidt did not take these precautions, so I am afraid his results at Boeing must be disregarded.

Quantum Cockroaches?

In the 1970s, Helmut Schmidt continued his experiments at the Mind Science Foundation in San Antonio, Texas, where he switched his focus from testing ESP to PK, the ability to influence events with the mind. In these experiments, subjects were asked to try to affect the output of a random number generator (RNG). Experiments were also conducted with animals and even cockroaches.

Other independent investigators have joined in this particular approach to psi experimentation. The 1987 National Research Council report mentions 332 separate RNG experiments between 1969 and 1984. Of these, 144 were conducted by the Engineering Anomalies Research Laboratory at Princeton University, under the direction of Robert Jahn, whose view of what constitutes an anomaly was discussed in an earlier chapter.

Princeton Anomalies

The Princeton laboratory greatly increased the sample of data on RNG experiments over the experiments of Schmidt. Jahn and his collaborators claimed deviations that could not be explained by chance (Jahn 1987). Did they succeed in replicating Schmidt?

Not quite. Scientific replication requires quantitative as well as qualitative agreement. Schmidt claimed a significant deviation from chance at about the one-in-a-hundred level. That is, his operators were able to affect the outcome of the RNG about one percent of the time. The Princeton success rate was considerably lower, less than one in a thousand (Druckman 1987).

This illustrates another property of psi experiments. As the techniques improve, employing tighter procedures and providing more data, the reported psi effects become smaller. Extrapolating to the ideal experiment with airtight procedures, we can expect zero effect.

In the Mood

Independent of statistical arguments, however, the Jahn experiments are marked by protocols that many scientists would regard as inadequate. The operator sits in front of the RNG, which displays both the individual trial scores and the accumulated mean of all trials so far generated. So she knows how she is doing as the series of trials proceeds. This feature is intended to provide a type of biofeedback, but it also makes possible a number of ways to produce a bias by other than paranormal means.

The laboratory supervisors "maintain a minimal presence during the actual operation," so the operator can feel comfortable and uninhibited. The experiment is carried out in a room with a "homey decor of panelled walls, carpeting, and comfortable furniture, with options for variable lighting, background music, and snacks." Furthermore, the operators are encouraged to engage in the experiments "whenever they are in the mood" (Jahn and Dunne 1987, p. 99).

As I have noted, psi researchers have always thought that psychic powers are depressed by too much supervision and the sterile environments of the typical research laboratory. Certainly nothing is wrong with providing people with more comfortable surroundings. However, we have seen that experiments with professional mediums were seriously flawed by the fact that the subjects themselves, rather than the experimenters, controlled the experiments.

Mediums characteristically insisted that their powers required just the right conditions. That may be, but it also gave them the opportunity to cheat, and based on the evidence, many did. Recall that even Rhine was disgusted by the obvious trickery of mediums. I wonder if he would have approved of the Princeton protocols.

Only with complete experimenter control and constant supervision can the prospect of cheating be adequately ruled out. In their experimental design, parapsychologists must take into account that skilled conjurers, even nonprofessional ones, can do some amazing things.

In the case of the Princeton experiments, one operator contributed to almost one quarter of the data and provided most of the significant results. Removing her scores reduces the remaining results to a level consistent with chance. Could this highly experienced operator, who was provided with immediate feedback on every trial, unwatched, and allowed to decide for herself the type of run to make, have found a way to bias the machine's output by nonparanormal means?

How? I can only speculate, and again I must make it clear that this is not an accusation of fraud, just a critical examination of the possibilities. Electronic circuits are known to "drift." They are often sensitive to heat, shock, and humidity. Perhaps the operator noticed a drift over the weeks, and took advantage of it. Perhaps she simply kicked the apparatus, turned it upside down, or blew on some of the transistors.

None of these may be the correct or even likely explanation of what occurred, but they illustrate that the experimental protocol was not sufficiently rigid. Remember, only a small systematic bias, one-tenth of one percent, is needed to produce the reported effect.

If the Princeton experimenters felt it was important that the operators be left alone, to conduct their tests when they were "in the mood," there still should have been other control trials where this was not the case. Furthermore, the tests should have been conducted under strict experimenter control, with baseline data (when no attempt is being made to mentally control the RNG output) taken at intervals uniformly interspersed with the intervals during which the operator tried to affect the outcome. These baseline intervals should not be too large, so that effects of normal instrument drift could be subtracted out. Judging by the reports, these precautions or other comparable ones were not taken.

The Verdict on RNG Experiments

The NRC study contained the following conclusion on the experiments by Schmidt, Jahn, and others that attempt psychic control of RNGs: "over a period of approximately 15 years of research, only one successful experiment can be found that appears to meet most of the minimal criteria of scientific acceptability, and that one experiment yielded results that are just marginally significant" (Druckman 1987).

The experiment referred to as "marginally significant" was conducted by Humphrey and Hubbard in 1980 and was not, at the time of the NRC report, published in a refereed scientific journal. It probably doesn't merit it. The chance probability for this experiment is an unimpressive 0.03, compared to numbers like one in a million we have seen in some earlier ESP reports.

A probability of 0.03 means that if 100 similar experiments are conducted, three, based on chance, would be expected to result in the same success rate as the reported experiment or better. Since a total of 332 RNG experiments were performed in the period considered, one three-percent effect is perfectly consistent with what one would expect from chance.

When Is a Result Significant?

The issue of when a scientific result is statistically significant is an important and highly misunderstood one. In medical research, new treatments with significance levels as low as three percent are usually acceptable, just in case they might help someone. In psychology, for similar reasons, presumably, five-percent effects are usually deemed adequate for publication.

In physics, by contrast, chance odds must be less than one in a thousand before anyone takes notice or gives serious thought to repeating the experiment. The best journals, such as *Physical Review Letters,* usually do not allow their authors to claim the discovery of a new phenomenon unless the chance odds are below one in ten thousand.

Parapsychologists have argued that the same criteria as used in the medical or behavioral sciences should apply to parapsychology, since it deals with people too. I would argue quite the opposite. Since parapsychologists are not in the business of helping people (at least directly and immediately), but rather join physicists in attempting to uncover fundamental new principles about the universe, the criteria for physics are more appropriate.

Ganzfeld Experiments

A skeptic who has gone further than most in giving parapsychology a fair reading is Ray Hyman, professor of psychology at the University of Oregon at Eugene. Hyman has met parapsychologists on their own turf, attending their meetings and writing in their journals.

As a founding fellow of the Committee for the Scientific Investigation into Claims of the Paranormal (CSICOP), Hyman has been a cautionary voice in skeptical circles, pleading the case for rational analysis instead of angry rhetoric. His influence has helped maintain the credibility of CSICOP as an organization willing to examine the facts and not pass judgment until the facts are clear. Still, Hyman's reasonableness has not prevented him from being attacked for his role as a member of the NRC's investigating committee (Alexander 1989).

In 1982, prior to his work for the NRC, Hyman agreed to do a careful critique of what was at that time a particularly promising avenue of psychic research, claiming replicable results: psi ganzfeld experiments. In a ganzfeld experiment, subjects are placed in a sensory-deprived state, with halves of ping pong balls taped over their eyes and white noise sound played in the background. Psi ganzfeld experiments are used to search for ESP on the reasonable assumption that it should be easier to achieve in the absence of normal sensory distractions.

The typical psi ganzfeld experiment is simple. Once the subject is in a relaxed state, a sending agent in another room proceeds to concentrate on a randomly selected picture. The subject, or percipient, then freely describes all the impressions he or she receives during the sending intervals.

Hyman was provided with the data from forty-two separate experiments involving forty-seven different investigators. Parapsychologist Charles Honorton, who had conducted five of the experiments, had previously classified twenty-three, or fifty-five percent, as having achieved a significance level of 0.05. That is, their individual chance probability was five percent or less. The chance probability of having twenty-three of forty-two experiments with this significance level is so low as to indicate, if the significance levels are calculated correctly, that some nonchance effect was almost certainly in operation.

Hyman's Critique

Hyman's report on his analysis of the ganzfeld studies was printed in the March 1985 issue of the *Journal of Parapsychology* (Hyman 1985). After correcting for biased reporting of the ganzfeld data, in which more successful than unsuccessful experiments are reported, he claimed that the significance level was really no more than an unimpressive thirty percent, not the reported five percent. In his report, Hyman also detailed many procedural flaws: inadequate randomization, potential sensory leakage, and statistical errors.

The same issue of the *Journal of Parapsychology* contained a lengthy rebuttal to Hyman's effort by Honorton (1985) and following issues contained further discussion of the matter, including a "Joint Communiqué" by Honorton and Hyman (Honorton and Hyman 1986). The abstract of the communique contains the sentence: "We agree that there is an overall significant effect in this data base that cannot be reasonably explained by selective reporting or multiple analysis." This was interpreted by some to mean that Hyman had backed off from his original conclusions and now saw evidence for psi in the ganzfeld experiments.

However, Hyman stuck by his conclusion that the experiments provided no such evidence. He later explained that the above quotation meant that selective reporting and multiple analysis cannot alone explain the effects reported. Other experimental defects explain the rest (Hyman 1989, p. 18).

In a personal letter to me, Hyman summarized his position as follows:

> The entire set of ganzfeld experiments do not justify drawing any conclusions about psi or anything else. They are plagued by too many departures from those ideals professed by the parapsychologists themselves. Given the distribution of statistical outcomes, it is unlikely that the results can be explained away as merely a statistical fluke or as the result of biased selection of the results. But even this statement should provide no comfort to the parapsychologists. This is because it is not even clear that the statistical outcomes are meaningful since the randomization procedures were inadequate. At any rate, because the test tube was clearly "dirty" in just about all of the ganzfeld experiments, these experiments cannot be used to support any hypothesis. (Hyman 1989)

The conclusion of the NRC on the ganzfeld studies was as follows: ". . . eight experiments were conducted with reasonable care, but none of these could be considered as methodologically ideal. When all 84 experiments are considered, it can be stated that the research methods are too weak to establish the existence of a paranormal phenomenon" (Druckman 1987).

Paraphysics

It has not been my purpose to do a complete review of a century and a half of psychic research. Ray Hyman has estimated that it would take someone five years full-time to do as complete a review as he did on the psi ganzfeld experiments on everything reported in recent years.

I have focused on the work of those individuals who have received the greatest notice within and without the psychic research community, such as William Crookes, Oliver Lodge, J. B. Rhine, and Helmut Schmidt. I am struck by the fact that so many parapsychological experiments over the years have been done by physicists rather than psychologists. One would think that psychologists would be more qualified to investigate matters of the mind. I guess the great majority of psychologists, who show no interest in getting involved in psi research, know what they are doing. Perhaps parapsychology should be renamed "paraphysics." In any case, the involvement of so many physicists allows me, as a physicist, to claim a certain qualification to comment.

Besides Helmut Schmidt, the two best known parapsychologists of recent years are also physicists: Russell Targ and Harold Puthoff. Their experiments in "remote viewing" with the Israeli magician Uri Geller and others, conducted at SRI International (formerly Stanford Research Institute) in Menlo Park, California, have received considerable attention and continue to be cited as scientific demonstrations of ESP.

Remote Viewing

The credibility of psi research obtained a big—if spurious—boost in October 1974, when the prestigious British journal *Nature* published Targ and Puthoff's paper on remote viewing (Targ and Puthoff 1974). One often hears parapsychologists and their supporters refer to "the research at Stanford published in *Nature*." They usually fail to note that SRI is not affiliated with Stanford University.

The *Nature* article was published only after some soul-seaching over several months by the editors. In an editorial in the same issue, they explained that three independent referees had agreed the paper was "weak in design and presentation, to the extent that details . . . were disconcertingly vague," revealing "a lack of skill" with "insufficient account of the established methodology of experimental psychology." The referees also were unanimous in the opinion that "the details given of various safeguards and precautions introduced against the possibility of conscious or unconscious fraud . . . were uncomfortably vague."

Then why publish it? If I had submitted a paper to *Nature* that received such dubious referee comments, it would have been rejected outright. Despite what the editors admitted were shortcomings that "on their own . . . could be grounds for rejection of the paper," they unwisely decided to go ahead anyway, so that the scientific community could pass judgment on a matter that had received considerable media attention. In no way, they insisted, does publication in *Nature* imply a seal of approval from the scientific community. However, this has not stopped others from claiming just that. In my opinion, *Nature* did the cause of rationality no favor by lowering its standards in this case.

Later, we will discuss the recent publication by *Nature* of an equally problematic article from a Paris medical laboratory claiming evidence that a certain antigen retained its effectiveness after being so highly diluted that not a single molecule was likely to remain (Davenas et al. 1988). We will see how the *Nature* editor then led a team, including magician James Randi, to investigate the procedures used, and then coauthored a follow-up article calling the experiments a "delusion" (Maddox et al. 1988).

J. B. Rhine had made an important contribution when he brought the investigation of psychic phenomena into the laboratory and performed experiments that provided, at least in principle, objectively measurable data. Although his experimental techniques were far from adequate, they still represented a vast improvement over the previous attempts at psychic research, which had relied on highly subjective anecdotes and unreliable eye-witness accounts and personal testimonials.

Unfortunately, Targ and Puthoff undid much of this progress, taking parapsychology a major step backward, reintroducing nonquantitative and subjective observations in their remote viewing experiments. These features, as we will see, doomed their efforts from the start.

The *Nature* article describes the experiments done with Uri Geller in an "electrically shielded room" where he reproduced pictures drawn by experimenters at remote locations, and a series of experiments with Pat Price describing remote outdoor scenes.

Superman

Uri Geller is the best-known and most financially successful professional psychic in recent times. He began his career as a stage magician in Israel, where he was once arrested for claiming that he performed his feats with psychic power. Apparently "truth in packaging" laws in other countries are not so strict as in tough-minded Israel, so Geller has since been able to make a tidy living for himself outside his home country.

Geller has now performed worldwide. His act consists of various standard mind-reading routines, such as guessing names of colors or foreign capitals written on a chalkboard behind his back. Such tricks are easily done with the help of an accomplice in the audience. He also reproduces pictures drawn by another person that are hidden from his view. Others have demonstrated this skill by following the top of the pencil. However, Geller is most known for his spoon-bending, which has been duplicated by other skilled magicians using distraction and sleight of hand. For complete critical reviews of the career of Uri Geller, see Randi (1975) and Marks and Kammann (1980).

Randi has personally duplicated all of Geller's best-known tricks. Marks and Kammann observed Geller during stage performances and other appearances in New Zealand. They detected an accomplice in the audience signaling Geller, and caught Geller bending keys physically (not psychically). They followed this up with their own experiments and were able to duplicate many of Geller's results with their students.

Geller at SRI

Targ and Puthoff originally invited Geller to SRI to test his supposed metal-bending powers. Most of his time there was occupied with such tests, but the results of the tests were never published. In the SRI experiment that was published in *Nature,* Geller was placed in various isolated rooms and asked to reproduce simple randomly selected line drawings. The *Nature* article describes thirteen trials, in eleven of which Geller provided drawings. The article actually reports only ten drawings, but later evidence showed that the experimenters discarded one because it did not match the target, claiming Geller had really declined that attempt because he was not getting any mental images.

The evidence also indicates that Geller declined three other attempts after doodling pictures, but Targ and Putoff included them anyway (Wilhelm 1976, p. 102). This kind of biased data selection violated basic principles of scientific methodology. Even worse, it is not mentioned in the article.

Both the target drawings and Geller's drawings are presented in the *Nature* article. Comparing them, one can see only one unambiguous case where Geller closely matched the target. A bunch of grapes is almost exactly reproduced, even to the position of the stem. Other Geller drawings are suggestive of the target drawing, but the correlation is not precise: a camel is drawn as a horse, a bridge becomes a banana, and a firecracker appears as a drum. Furthermore, Geller often provides several different drawings to chose from, improving the odds that one will be acceptable

Nevertheless, two independent judges were able to make a one-to-one match of Geller's ten drawings with the originals. The probability for this to happen randomly with ten independent pairs of objects is three in ten million.

The same month that the *Nature* article by Targ and Puthoff appeared, the British magazine *New Scientist* published a long article on the results of its investigations of Geller phenomena by Dr. Joseph Hanlon. The magazine had set up its own series of tests that Geller had originally agreed to participate in, but then backed down. Hanlon describes in detail how Geller's many reported feats could have been done by magician tricks. Only the SRI experiments with Geller came close to providing sufficient controls against trickery, and even these failed in Hanlon's view (Hanlon 1974).

Hanlon suggested that Geller could have received signals in the SRI experiments by a miniature radio. Targ and Puthoff scoffed at this, saying it was impossible. One of Geller's drawings, one of the sun, was produced in a Faraday cage, a copper mesh designed to electromagnetically isolate its volume, but this was the only drawing that was reported under such conditions.

Later investigations by James Randi and other skeptics pointed to many other defects in the experimental procedure that would have made it possible for a skilled conjurer like Geller, aided by accomplices, who were known to be in the vicinity and not carefully watched, to have obtained sufficient information to produce crude sketches resembling the original drawings (Randi 1975, Wilhelm 1976, Marks and Kamman 1980).

For example, the steel room used in six of the trials was not totally closed off. A hole was cut in it for electrical cables and a window was covered with a bulletin board that Geller could easily have cut a tiny peephole in, through which he could have been signaled. The room also had a two-way intercom that might have been surreptitiously used by an accomplice. The Faraday cage used in two trials was transparent.

Marks and Kammann provide what they call a "feasible script" for the thirteen Geller trials (Marks 1980). Little can be proved because of the inadequate documentation and lack of cooperation provided by SRI to those who have tried to dig further into the Geller experiments. But since Geller has been caught numerous times using trickery in other contexts, and since the opportunity clearly existed for him to have obtained information by nonpsychic means at SRI, the Geller experiments at SRI must be disregarded as evidence for psychic power. Once again, we are forced by Occam's razor to reject the paranormal explanation in the light of more likely normal ones.

The Pat Price Experiment at SRI

The other remote viewing experiment reported by Targ and Puthoff in *Nature* was conducted differently, using natural targets instead of line drawings. The subject remained in the laboratory while a target team visited a randomly selected location. The subject was then asked to describe his impressions of the site by drawing pictures and talking into a tape recorder.

The first nine experiments were performed with subject Pat Price, an ex-police commissioner and vice-mayor of Burbank, California, who claimed to have had great success using his psychic powers to detect the locations of criminals. Sites in the Bay Area were randomly selected by a double-blind protocol. The sites included the Hoover Tower, a radio telescope, and a drive-in theater.

Five judges were taken to each of the sites and asked to match them with the subject's drawings and verbal descriptions on tape. Of the forty-five selections (five judges, nine selections each), twenty-four were correct. The chance probability was calculated to be 8×10^{-10} (Targ and Puthoff 1974). In their book *Mind-Reach,* Targ and Puthoff claimed to have performed over a hundred other tests, most of which were successful (1977).

The Attempt to Duplicate

Inspired by the astounding success and unique features of the SRI technique, New Zealand psychologists David Marks and Richard Kammann attempted to duplicate the experiment. The results are reported in their book *The Psychology of the Psychic* (1980).

Marks and Kammann performed a total of thirty-five tests with five different subjects and five judges. None of the results was statistically significant.

The one important difference between the Marks-Kammann and Targ-Puthoff experimental protocols was that Marks and Kammann carefully edited the transcripts containing the subjects' descriptions of the images that came to their minds, so that they contained no references to previous targets.

Marks and Kammann Shoot Down Targ and Puthoff

In 1978 Marks visited SRI and was allowed to study the Price series transcripts. He found a number of clues and hints that could have helped the judges decide which site should be matched with a given description. For example, references are made to an experiment being "the second of the day." Targ is heard to say, "Nothing like having two successes behind you" and mentions a site visited the day before. Using these and other clues, Marks was able to match all of an earlier agreed-upon five descriptions to their sites, without visiting the sites. These results, totally invalidating the SRI experiment, were also published by *Nature* (Marks 1978).

The Verdict on Viewing

The report of the NRC summarizes the situation with respect to remote viewing. The authors expressed surprise that, given the enormous publicity and unusually strong claims for remote viewing, only twenty-eight formal experiments had been performed over the ten-year period reviewed. Of these, only thirteen claimed positive results, a failure rate far above what would be expected on the basis of Targ and Puthoff's statement that they succeeded with every subject they ever tried. Further, only nine of the supposedly successful experiments were published by refereed journals. Seven of the nine experiments had been conducted by Targ and Puthoff. The NRC examined these nine and found each to be seriously flawed.

The methodological flaw found in all but one experiment was the lack of independence between successive trials. The NRC report goes into this flaw in great detail, but it relates basically to the sensory clueing discovered by Marks and Kammann. In the single experiment (of the original twenty-eight) where a positive effect is reported and sensory clueing can be reasonably ruled out, a member of the target team familiar with the remote targets translated the subject's descriptions into Italian for the judging process. The NRC remarks: "Why the experimenters allowed such potential sources of biased experimental procedures is not known, but the violation obviously negates the results as evidence for psi."

The committee's conclusion on remote viewing: "After approximately 15 years of claims and sometimes bitter controversy, the literature on remote viewing has managed to produce only one possibly successful experiment that is not seriously flawed in its methodology—and that one experiment provides only marginal evidence for the existence of ESP. By both scientific and parapsychological standards, then, the case for remote viewing is not just very weak, but virtually nonexistent" (Druckman 1987).

Wanting To Be Fooled

We saw earlier that Uri Geller had succeeded in pulling the wool over the eyes of scientists Lyall Watson and John Taylor in England (Taylor 1980, Watson 1973). Taylor eventually realized what had happened. He had simply been fooled, and was professional enough to admit it.

Scientists are not trained in the detection of trickery. Quarks and leptons may be elusive, but they never lie to you. In the cases of Watson, Tart, Targ, and Puthoff, however, we see examples of scientists evidently asking to be fooled, so committed are they to the reality of the phenomenon—despite the evidence. I have great difficulty believing that trained, honest scientists with all their marbles can really be that gullible.

Why ask to be fooled? I can think of two reasons. First, a scientist might be willingly fooled because of some deep-seated religious convictions and belief in the supernatural. This will to believe is sincere, though misguided. William Crookes, Oliver Lodge, and, I think, J. B. Rhine, fit into this category.

Alternatively, the scientists might be less sincere, and see in psychic trickery a way to gain personal fame and fortune. In our society, ample opportunity exists for fortunes to be made by someone who can capture the market with some new psychic mumbo jumbo.

Parapsychology: A Pseudoscience

Today, parapsychology is widely regarded as a pseudoscience, not because psychic phenomena remain unverified—though this is the case—but because for over a century it has been tainted by fraud, incompetence, and a general unwillingness to accept the verdict of conventional scientific method.

In their books and articles, parapsychologists claim a commitment to the standard precepts of scientific method. When I read their principles of methodology, I find little to disagree with. They say most of the right things about careful controls, double-blind protocols, and independent replication.

But investigators, like politicians, should be judged by what they do rather than what they say. The experiments we have considered, which by the parapsychologists' own reckoning are the best they have been able to produce in a century and a half, are simply poor science.

A certain amount of floundering was acceptable during the early years of psychic studies, while people were still formulating their ideas and methods. We should not be too critical of the serious investigators who, a century ago, tried to come to rational terms with the widespread reports of supernatural phenomena. However, after this length of time, contemporary parapsychologists should stop beating the dead psychic horse and take up other more profitable lines of work. If ESP existed, it surely would have been found by now.

A Popular Old Idea

The record speaks for itself of the willingness of scientific leaders, such as Einstein, to let the psychic researchers have their shot. They have had it. Still, however, parapsychologists seek public sympathy by claiming that they are prevented by the scientific establishment from entering into the science club with their "new ideas." But in the case of psychic phenomena we are not dealing with some unpopular new idea.

The existence of supernatural forces is rather a popular old idea, as old or older than history. As D. H. Rawcliffe has put it, "psychical research, or parapsychology, has its genesis in the myths, the folklore, the magic and mysticism of pre-civilization. . . . There is . . . an unbroken historical continuity observable between superstitious occultism and the more sophisticated occultism which flourishes at the present time" (Rawcliffe 1959).

Even the most modern form of secular supernaturalism, psychic phenomena, has been promoted for over a century. So we skeptics are not resisting some brilliant young Galileo or Einstein who has just made a startling new discovery by looking through his telescope or working through his equations. Rather we question the antiquated notions of mind beyond matter, spirit, soul, and astral planes which, though constantly re-coated in the latest fashionable veneers, are really nothing more than the rickety termite-eaten furniture of three-thousand-year-old superstitious beliefs.

References

Alexander, Col. John. March/April 1989. "A Challenge to the Report." *New Realities*. P. 10.

Davenas, E., et al. 1988. "Human Basophil Degranulation Triggered by Very Dilute Antiserum against IgE." *Nature* 330:816-818.

Druckman, Daniel, and Swets, John A., eds. 1987. *Enhancing Human Performance: Issues, Theories and Techniques.* Washington, D.C.: National Academy Press.

Eyseneck, Hans J., and Sargent, Carl. 1982. *Explaining the Unexplained.* London: Weidenfeld and Nicholson.

Frazier, Kendrick, ed. 1981. *Paranormal Borderlands of Science.* Buffalo, N.Y.: Prometheus Books.

———. 1986. *Science Confronts the Paranormal.* Buffalo, N.Y.: Prometheus Books.

Gardner, Martin. 1957. *Fads and Fallacies in the Name of Science.* New York: Dover.

———. 1981. *Science: Good, Bad and Bogus.* New York: Avon.

Hanlon, Joseph. 1974. "Uri Geller and Science." *New Scientist* 64:170-185.

Hansel, C. E. M. 1966. *ESP: A Scientific Evaluation.* New York: Scribner's.

———. 1980. *ESP and Parapsychology: A Critical Reevaluation.* Buffalo, N.Y.: Prometheus Books.

———. 1989. *The Search for Psychic Power: ESP and Parapsychology Revisited.* Buffalo, N.Y.: Prometheus Books.

Hines, Terence. 1988. *Pseudoscience and the Paranormal: A Critical Examination of the Evidence.* Buffalo, N.Y.: Prometheus Books.

Honorton, C. 1985. "Meta-analysis of Psi Ganzfeld Research: A Response to Hyman." *Journal of Parapsychology* 49:51-91.

Honorton, C., and Hyman, Ray. 1986. "A Joint Communiqué: The Psi Ganzfeld Controversy." *Journal of Parapsychology* 409:351-364.

Hyman, Ray. 1985. "The Ganzfeld Psi Experiment: A Critical Appraisal." *Journal of Parapsychology* 49:3-49.

———. 1989. *The Elusive Quarry. A Scientific Appraisal of Psychical Research.* Buffalo, N.Y.: Prometheus Books.

———. 1989. Private communication.

Jahn, Robert J., and Dunne, Brenda J. 1987. *Margins of Reality: The Role of Consciousness in the Physical World.* New York: Harcourt Brace Jovanovich.

Kennedy, J. L. 1939. *Psychological Bulletin.* P. 91.

Kurtz, Paul. 1986. *The Transcendental Temptation: A Critique of Religion and the Paranormal.* Buffalo, N.Y.: Prometheus Books.

Maddox, John, Randi, James, and Stewart, Walter W. 1988. " 'High-Dilution' Experiments a Delusion." *Nature* 334:287-290.

Marks, David, and Kammann, Richard. 1978. "Information transmission in remote viewing experiments," *Nature* 274:680-681.

———. 1980. *The Psychology of the Psychic.* Buffalo, N.Y.: Prometheus Books.

Markwick, Betty. 1978. "The Soal-Goldney Experiments with Basil Shackleton: New Evidence for Data Manipulation." *Proceedings of the Society for Psychical Research* 56:250-277.

Mauskopf, Seymour H., and McVaugh, Michael R. 1980. *The Elusive Science: Origins of Experimental Psychical Research.* Baltimore, Md.: Johns Hopkins University Press.

Medhurst, R. G. 1971. "The Origin of the 'Prepared Random Numbers' Used in the Shackleton Experiments." *Journal of the Society for Psychical Research* 56:43-72.

Palmer, John A., Honorton, Charles, and Utts, Jessica. 1989. "Reply to the National Research Council Study on Parapsychology." *Journal of the American Society for Psychical Research* 83:31.

Pratt, J. G., and Woodruff, J. L. 1939. "Size of Stimulus Symbols in Extra-Sensory Perception." *Journal of Parapsychology* 32:121-158.

Price, George R. 1955. "Science and the Supernatural." *Science* 122:359-367.

Randi, James. 1975. *The Magic of Uri Geller.* New York: Ballantine Books.

Rawcliffe, D. H. 1959. *Illusions and Delusions of the Supernatural and the Occult.* New York: Dover.

Rhine, J. B. 1929. "An Investigation of a Mind Reading Horse." *Journal of Abnormal and Social Psychology* 23:449-466.

———. 1934. *Extrasensory Perception.* Boston: Bruce Humphries.

———. 1954. *New World of the Mind.* London: Faber & Faber.

Rhine, J. B., and Pratt, J. G. 1954. "A Review of the Pearce-Pratt Distance Series of ESP Tests." *Journal of Parapsychology* 18:165-177.

Schmidt, Helmut. October 1969. "Quantum Processes Predicted?" *New Scientist.* Pp. 114-115.

Scott, Christopher, and Haskell, Philip. 1974. "Fresh Light on the Shackleton Experiments." *Journal of the Society for Psychical Research* 46:43-72.

Soal, S. G., and Bateman, F. 1954. *Modern Experiments in Telepathy.* London: Faber & Faber.

Soal, S. G., and Goldney, K. M. 1943. "Experiments in Precognitive Telepathy." *Proceedings of the Society for Psychical Research* 47:21-150.

Smith, W. R., Dagle, Everest F., Hill, Margaret D., and Smith, John Mott. May 1963. *Testing for Extra-Sensory Perception with a Machine.* Data Sciences Laboratory Project 4010.

Targ, Russell, and Puthoff, Harold. 1974. "Information Transmission under Conditions of Sensory Shielding." *Nature* 251:602-607.

———. 1977. *Mind-Reach.* New York: Delacorte Press.

Targ, Russell, and Harary, K. 1984. *The Mind Race: Understanding and Using Psychic Abilities.* New York: Villard.

Tart, Charles. 1976. *Learning to Use Extrasensory Perception.* Chicago: University of Chicago Press.

Taylor, John. 1980. *Science and the Supernatural.* New York: Dutton.

Thomson, Joseph John. 1936. *Recollections and Reflections.* London: G. Bell.

Thouless, R. H. 1974. "Some Comments on 'Fresh Light on the Shackleton Experiments.' " *Proceedings of the Society for Psychical Research* 56:88-92.

Watson, Lyall. 1973. *Supernature.* London: Hodder & Stoughton.

Wilhelm, J. L. 1976. *The Search for Superman.* New York: Pocket Books.

Zusne, Leonard, and Jones, Warren H. 1982. *Anomalistic Psychology: A Study of Extraordinary Phenomena of Behavior and Experience.* Hillsdale, N.J.: Lawrence Erlbaum Associates.

9.

Harmonica Virgins

One of the proofs of the immortality of the soul is that myriads have believed it—they also believed the world was flat.

Mark Twain

Galactic-Solar Songs Psychically Received

In August 1987, handfuls of New Age faithful, accompanied by greater numbers of the curious, gathered at certain sacred sites around the country where resonances of vibrations of the cosmic fluid were expected to be amplified and focused. This remarkable event was called the "Harmonic Convergence." One of the prime sites was Diamond Head crater, the famous Waikiki landmark. A small group, dubbed "Harmonica Virgins" by a local wag, assembled there just before dawn on Sunday, August 16.

Although no diamonds have ever been seen in Diamond Head except on the fingers of affluent tourists, the mere suggestion of crystalline structure in the popular name for Mount Leahi was apparently sufficient to focus cosmic energies on the long extinct volcano. No detectable effects were reported, but the harmonica virgins went home happy. To the New Ager, everything—even death—is a "growth experience."

Harmonic Convergence was the brainchild of art historian José Arguelles. In his book *The Mayan Factor: The Path Beyond Technology,* Arguelles predicted that a twenty-five-year cycle would commence in 1987 and culminate in 2012 with the "closing out . . . of the evolutionary interim called homo sapiens. Amidst festive preparation and awesome galactic-solar songs psychically received, the human race, in harmony with the animal and other kingdoms and taking its rightful place in the great electromagnetic sea, will unify in a single circuit. . . ." (Arguelles 1987).

The New Age

The hodgepodge of occult ideas and empty scientific jargon called the New Age is the latest variation on a continuing theme that we hear replayed throughout history. Forms of supernaturalism not associated with any particular church have popped up in countless varieties since earliest times, from ancient astrology and alchemy to more recent spiritualism and psychicism.

As the latest reincarnation of secular supernaturalism, the New Age has collected ideas from just about every mystical and occult movement that ever occurred in the East or West, discarding very few, if any. According to Christopher Lasch, it has combined "meditation, positive thinking, faith healing, rolfing, dietary reform, environmentalism, mysticism, yoga, water cures, acupuncture, incense, astrology, Jungian psychology, biofeedback, extrasensory perception, spiritualism, vegetarianism, organic gardening, the theory of evolution, Reichian sex therapy, ancient mythologies, archaic nature cults, Sufism, Freemasonry, Cabalistic lore, Chiropractic, herbal medicine, hypnosis, and any number of other techniques designed to heighten awareness, including elements from the major religious traditions" (Lasch 1987, p. 81).

To the above list can be added: karma, reincarnation, past-life regression, near-death experiences, out-of-body experiences, astral travel, homeopathy, naturopathy, firewalking, UFOs, Atlantis, Lemuria, pyramid power, tarot, the Bermuda Triangle, ancient astronauts, Gaia, the Shroud of Turin, holism, quantum mechanics, Transpersonal Psychology, Swedenborgianism, Christian Science, Theosophy, Anthroposophy, Scientology, TM, Eckankar, EST, the Trancendentalism of Ralph Waldo Emerson, and even the Akashik records of James Joyce's *Ulysses*—containing "all that ever anywhere wherever was."

All this is blended, together with large doses of pseudoscience, to make a brew that, in its surface form, appears to represent a composite of modern spiritual thinking. Thus the *atman, nirvana,* and *satori* of Eastern mysticism have been transformed into "universal energy" and "morphic resonances." And modern physics is called upon to testify that the universe is one unbroken whole, that every complex system from atoms and crystals to human societies, the earth, and all the cosmos have minds. Further, the human mind is linked to each of these other minds in one great cosmic consciousness.

The New Age dates back to the 1970s, when the antiwar and hippie movements flourished and developed mystical components. In those days, Harvard professor Timothy Leary and others were experimenting with psychedelic drugs such as LSD, claiming that they opened the mind to new levels of reality. One of Leary's colleagues was a psychology professor, Richard Alpert. Alpert, born Jewish, converted to Eastern mysticism and, under the new name Baba Ram Dass, wrote a series of popular books that form the theoretical basis for the New Age (Dass 1973, 1977, 1978, 1979).

Equally as influential as Dass in shaping the New Age movement has been Marilyn Ferguson, who has written numerous essays and a best-selling book, *The Aquarian Conspiracy* (Ferguson 1980). The "conspiracy" that Ferguson speaks of is far from an evil one. Rather it is "a leaderless but powerful network . . . working to bring radical change." The change, in Ferguson's view, would be for the better.

Ferguson says her New Age network has broken with much of Western thought because science has placed limits on human potential. The New Age, on the other hand, reaffirms the ancient belief that the human psyche is limitless—transcending space, time, and matter. To support this, Ferguson says that psychic and occult phenomena "irrefutably occur" and can be "facilitated by psychotechnologies." The result is a new paradigm in which human beings have the power to produce miracles, and even generate new realities.

Falling Off a Limb

Transcending both Dass and Ferguson as chief public spokesperson for the New Age movement has been the Academy Award–winning actress Shirley MacLaine. Starting with a bestseller in 1983 called *Out on a Limb* (MacLaine 1983), MacLaine has written a series of books that include *Dancing in the Light* (1985) and *Its All in the Playing* (1987). They have sold millions. She has also starred in a TV production centered on her mystical experiences, appeared as a guest on countless talk shows, and traveled the nation giving $300 per person seminars on the new awareness.

MacLaine's theme is that we can do no wrong: "evil is live spelled backwards"; that we create our own reality: "I'm creating you right here"; that we each are God: "I am God." She believes that crystals have special psychic powers: "When you hold crystals, they amplify thought waves"; and that energies are associated with consciousness: "When three or more people are gathered with the same intentionality in a room, the energy units are squared." MacLaine sees a millennium ahead: "The vibrational oscillation of nature is quickening," presumably by utilizing her psychic powers: "In deja vu you are getting an overlap of a past-life experience, or you could be getting an overlap of a future-life experience. . . . That's what Einstein said."

Need I go further? For critical analyses of Shirley MacLaine's ideas, see Gordon (1988) and Gardner (in Basil 1988, p. 185). For skeptical views of the New Age movement in general, see the book by Martin Gardner (1989) and the collection of essays edited by Robert Basil (1988).

In a later chapter I will examine some of the philosophical issues that have been raised by more sophisticated New Age thinkers. There I adopt the optimistic attitude of the child in one of former President Ronald Reagan's favorite funny stories, who wakes up Christmas morning to find nothing but a pile of horse manure in the middle of the living room. Grabbing a shovel he digs in, figuring that "there has to be a pony in there somewhere."

Getting into the Game

Certainly the notion of psychic powers remains the keystone of the New Age, just as it was of the older traditional spiritualism, secular and religious. I have surveyed the attempts that were made over the past century and a half to apply scientific techniques to the search for evidence of psi. We have seen that such evidence has not yet been found, and I have tried to make it clear why we are forced to that conclusion, despite many vocal claims to the contrary.

But so far I have limited the discussion to a critical analysis of other people's efforts. I have played devil's advocate, or skeptic, a position that must be filled when the game of science is played. Yet, this position is on the sidelines, and it seems all the devil's advocate has to do is boo the players who foul up on the field. But you can't understand baseball until you have been up at the plate yourself to try to make contact with a ninety-mile-per-hour fastball. So let me tell you about some of my own modest efforts to get into the game of psychic research.

Skeptical Convergence

The presence of the Diamond Head focal point for the 1987 Harmonic Convergence just a few miles from my office on the Manoa campus of the University of Hawaii inspired a group of us to have our own convergence—of skeptics. So, a few weeks later, suitably on Halloween, the "Hawaii Skeptics" sponsored a special event we called "Skeptical Convergence."

The ballroom on the Manoa campus was rented, and a two-day admission-free fair was held that was well attended by students, members of the community, and the media. The local newpapers gave it front-page coverage, and all three major TV stations did news reports, one of which included a live interview with co-organizer Kurt Butler and myself. Clearly we had hit upon an effective means, through the entertaining medium of a fair, to get our skeptical message transmitted to the public.

While we did not focus particularly on the New Age, many aspects of that movement were represented in the program. Displays provided information on varieties of paranormal claims and questionable health practices, all promoted in New Age thinking, from UFOs to Chiropractic. Cash prizes were offered to anyone who could levitate sufficiently to reduce his or her weight by ounces, or read a poker hand inside a safe. There were no winners.

The "Shroud of Manoa," with stains looking remarkably like a happy face, was on display. Astoundingly accurate astrological personality readings were given (suitably randomized). A videotape was played of a group of us walking on red-hot coals. The Skeptics' firewalking demonstrated that nothing violating any principles of physics is involved in this phenomenon, and no costly "consciousness-raising" seminar is involved: You simply wet your feet and run.

Many youngsters attended the Skeptical Convergence, and they particularly enjoyed lying on the Bed-of-Nails, another supposed mind-over-matter feat that is simple physics. I had borrowed the bed from a physics graduate student who used it in his classes to demonstrate principles of pressure and force.

Rational skepticism does not rule out the acceptance of well-established scientific claims, and we presented information on the dental value of fluoridation, with a taste test to allow people to find out for themselves whether they can detect the presence of fluoride in water.

A professor of religion at the University of Hawaii demonstrated street magic of India. From among the many feats he had learned from his research in India, he punctured his arm with needles so that blood flowed. He allowed the audience to examine his wounds close-up. Then he miraculously healed his wounds, so that no sign of them remained.

Another magician was able to read the mind of a young lady in the front row whom he had never met, accurately describing her apartment, and incredibly intuiting her vegetarianism and devotion to animals. Now it happened that this woman was a good friend of my wife and me, but of course no one entertained the thought that I, a scientist renowned for my integrity, might stoop so low as to have fed the magician information beforehand.

Later, after the magician explained that everything was done by trickery (though, following the rules of his profession, he did not explain how), some people in the audience still insisted he had psychic powers.

Converging on ESP

As one of my displays, I prepared a computer-controlled ESP test that was open to all participants in the fair. Although I have already dealt extensively with experiments on ESP, allow me to present some of the details of my personal experiment. They turned out to illuminate a number of fundamental issues.

My test was based on the classical Rhine ESP card-guessing game. Subjects were given twenty-five chances to use whatever powers they possessed to determine which of the five ESP symbols displayed on a computer screen had been randomly selected by the computer program.

Though you might argue that telepathy cannot occur when an inhuman computer brain is doing the selecting, clairvoyance or precognition would presumably still work. Or, as with the Schmidt and Jahn experiments, perhaps psychokinesis could be used to move the electrons through the computer circuits. A Heath/Zenith HS-151 Personal Computer was used, with the MS-DOS BASIC random number generator selecting the targets.

The five symbols were presented on the computer screen as shown in Figure 9.1. These are the basic Zener symbols of the ESP cards used in the Rhine experiments, except that three vertical straight lines were used instead of wavy lines since they were easier for me to program.

Figure 9.1. The symbols used in the author's ESP experiment. They are identical to the Zener or ESP symbols used in early card-guessing experiments, except that the fifth figure is composed of straight lines rather than wavy ones.

The Problem of Subject Control

In many parapsychological experiments, subjects are given immediate feedback, and are often also given the option to quit on their own volition when they see they are not doing well, or to continue longer than scheduled when they "are on a roll." This procedure is based on various unsubstantiated properties of ESP such as the "decline effect" that Rhine and other para-psychologists claimed to detect, and their general belief that ESP abilities can be trained. But, as we have seen, there is insufficient evidence that ESP even exists. So no one can claim that any special properties have been demonstrated.

The problem with the subject controlling the duration of an experimental run is that it produces a statistical bias in the data very difficult to correct for after the fact. To avoid this problem, my subjects were not given their scores until the end of their sessions, so as not to discourage them from completing all twenty-five trials. Further, I did not accept any data from people who did not complete all twenty-five trials and allowed no more than twenty-five trials for a given test of a subject.

Looking for One-in-a-Thousand

Beforehand I had programmed the computer to ring a bell for a score of twelve correct hits or better, in order to invite the subject back in this case for further testing. This twelve hits criterion corresponded to a chance probability of less than one-in-a-thousand. That is, I would expect, on average, fewer than one person in a thousand to be able to make twelve or more correct guesses.

I made no claim to try to "prove" or "disprove" ESP by this trivial experiment. My declared intention was to simply test the hypothesis that ESP is a common characteristic in the general population, an assertion often made by its promoters, and provide a challenge for all who wished to test their powers.

I readily admit that more sophisticated experiments have been done to test ESP, and my purpose for discussing this one is not to suggest that it provides any startling new results. However, I think the results are probably typical, and so worth providing as an example.

In particular, my experimental procedure demonstrates a number of the problems and pitfalls of this type of experimentation. We can see how some people who are untrained in statistics and experimental techniques can easily find what they think is evidence for ESP. Many courses and lecture demonstrations of ESP use similar simple tests, and the students or audience often go away convinced that they have the power.

Eighty Honest Citizens

The experiment tested the powers of eighty subjects. Each made twenty-five trials, so that the two-day experiment comprised 2,000 trials. While

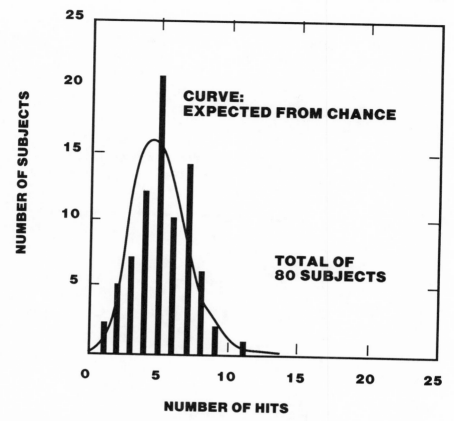

Figure 9.2. Distribution of hits in the author's ESP experiment. The solid curve indicates what is expected from chance. Note that the data are skewed slightly higher than the curve. One subject scored 11 of 25 possible hits, for a chance probability of 0.15%.

this may not seem like many, eighty honest citizens are still better than one sleight-of-hand artist. Furthermore, I obtained some interesting results.

In Figure 9.2, the actual numbers of hits (bar graph) are compared with the expected numbers based on pure chance (solid curve). Most scientists seeing this result would say that the data are consistent with the theoretical curve, that is, with chance. But a few things are notable.

First, there appears to be a slight upward skewing of the actual results, suggesting a small positive effect. Second, one subject scored eleven hits, just below my bell-ringing threshold where the subject would be invited back for further testing. Despite this subject's excellent results, I did not invite him back, since that would have changed my experimental protocol after the experiment was performed, an unacceptable scientific procedure often followed in psi experiments.

The chance odds for eleven or more hits in twenty-five trials for five objects is 1/667; that is, if the experiment were done with 667 subjects, I would have expected one subject to get at least eleven hits by chance. Yet there weren't 667 subjects, only eighty. So the one score of eleven was an unlikely result. But is it significant?

Mystical Squares and Circles

Before answering, let me note another interesting result that also bears on the question of significance. Figure 9.3 shows a plot of the number

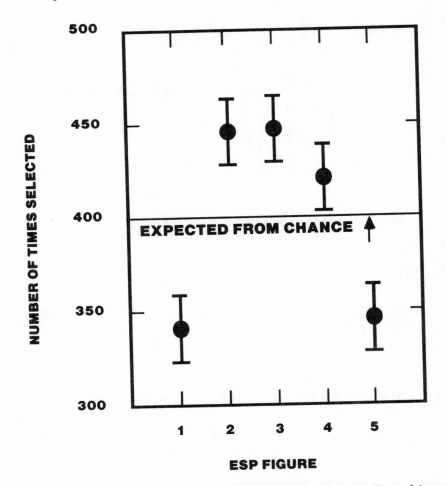

Figure 9.3. The number of times each of the five figures shown in Figure 9.1 was chosen in the author's ESP experiment. The line at 400 indicates the number expected by chance. The bars on the data points indicate the range of fluctuation expected for experiments with the same number of trials. The odds that the observed results were due to chance are one in 40,000.

of times the subjects chose each ESP symbol; the numbers 1-5 correspond to the shapes shown left-to-right on the screen, as in Figure 9.1.

In Figure 9.3, I have presented the experimental results according to a typical method used to present physics data, with "error bars" on the data points indicating the statistical standard deviation, that is, the range of random fluctuation that one might expect to see for each choice in a large number of identical experiments. The standard deviation, or sigma, is essentially the square root of the expected number of choices in this case. Since each symbol was expected to be chosen 400 times on average (2,000 total trials divided by five symbols), the standard deviation is plus or minus twenty.

So if all five symbols are chosen by the subjects with equal preference, we would expect each symbol to be chosen 400 times, give or take about 20. For example, if the results of the experiment had been, hypothetically, 390, 405, 393, 417, 395, respectively, for the five symbols, the results would have been consistent with chance.

But this is not the way it turned out. Instead we see that the first symbol (the cross) was only picked 340 times, sixty times less than expected. The fifth symbol (the lines) was almost equally unpopular. By comparison, the middle three symbols were preferred, with the second symbol (the square) and the third symbol (the circle) each chosen at least forty times more frequently than would be expected if all five selections were equally likely. The odds against the observed data resulting by chance is 40,000 to one. Some great mystical significance must attach to squares and circles!

The Brain is a Poor Random Number Generator

Have I discovered ESP in a simple two-day experiment? As I keep telling people who suspect my motives, I would love to do so. Surely great fame and fortune awaits the scientist who finally convinces the rest of the scientific community that a world beyond the normal senses exists, and that this world is accessible to the human mind. The only reason I could possibly have to hide a positive result is that I personally possess ESP power and wish to keep it hidden from the world so that I might use it to my advantage. Ridiculous as that sounds, it has been suggested as the motive for my skeptical activities!

Unfortunately, the significant deviation from chance I am reporting here is no signal for ESP. After mulling over my results, I now see that I introduced a bias by my placement of the symbols on the screen. The subjects' attention was probably drawn to the prominent circle and square in the center. I should have put these at the ends; though perhaps the central symbols would have been preferred no matter what they were. Or, maybe the circle and square would have been preferred no matter where they were. I might have randomized the placement, but then, wouldn't individuals who randomly saw the circles and squares at the center score more hits? And wouldn't Christians tend to choose the cross more frequently? Or perhaps avoid it because of unpleasant connotations? Who knows?

This illustrates one of the problems with any experiment involving human subjects: people are poor random number generators, and so you cannot simply compare their guesses with chance effects, and then assume you have made some great discovery when the expectations of chance are not confirmed.

Yet this is precisely what is done in many of the ESP experimental tests that claim statistically significant positive effects. For example, Helmut Schmidt's highly touted results are reported to have odds of 100 million to one against having occurred by chance. So what? If I kept conducting my biased experiment with more subjects and trials, I could easily achieve 100 million to one against chance.

For example, just one additional day with forty more subjects would have given me another 1,000 trials. If they were affected by the same bias as those of the first two days, odds against chance of 20-million-to-one would be achieved for the three-day experiment. The odds after a fourth day would be astronomical. And what would it prove? Simply that people do not select ESP symbols with any greater impartiality than they exhibit in selecting their spouses or political leaders.

The Lesson

If ESP is ever to be demonstrated, it will have to be with an unambiguous prediction of previously unknown events, a test that natural sciences meet regularly. As mentioned, I did find a slight indication of excess hits in my experiment. Since the computer by chance happened to choose the square somewhat more frequently in this sequence of tests, and the square was a popular choice of the subjects, a slight upward skewing of the hit distribution is expected—but for natural, not paranormal, reasons.

I have no idea how one can properly estimate the probability that these results are consistent with whatever nonrandom mental algorithms were being executed in the brains of the subjects when they made their selections. And not being able to do that, how could one ever convincingly demonstrate a paranormal ESP effect in any experiment that compares results with chance? Calculating the natural probability from chance to find evidence for a paranormal effect is clearly unproductive, as the long unsuccessful history of experiments of this type testifies. The brain does not work that way at all.

A much better way to test for ESP is to have the computer make its selections according to some equation or algorithm selected by some disinterested outsider and kept secret from subject and experimenter alike. Then see if any subject is able to reproduce the algorithm. This double-blind protocol is surprisingly absent from many psi studies.

Is Something There?

So the upward skewing of results in my experiment means nothing, and the fact that the distribution of selected symbols has only a minute probability of being chance is certainly no indication of ESP. What about the fellow who scored eleven times out of twenty-five? Is this yet another case—like so many reported in the parapsychological literature during 130 years of psychic research—wherein a marginal effect added to so many other marginal effects is thought to provide a convincing case that "something is there"?

All I can say is that, in my experience, I have seen any number of reported effects with a chance probability as low as 1/667 fail to be replicated by further independent investigation. Chance effects at this level will happen on average of once every 667 experiments, and thousands of experiments are done every year in each major branch of science.

A Thousand To One Means Nothing

In an experiment done on the top of Mount Hopkins in Arizona, my collaborators and I once thought we had discovered a new trillion electron-volt source of gamma rays in the cosmos. It seemed very convincing, with single-trial odds for a chance effect less than one in a billion. Naturally we became very excited and had dreams of glory.

However, before submitting the result for publication, we very carefully surveyed the data analysis procedures that had been followed. To our surprise, we realized that we had actually made a million separate independent trials on the computer in searching for the source. In calculating chance probabilities, one must multiply the single-trial probability by the number of trials. Thus, instead of a billion to one, our actual odds against chance were an unimpressive thousand to one. Statistical flukes at this significance level will happen about once in every thousand experiments. So we never sent the paper in for publication. Since then, our source has not been reported by anyone else. Like ESP, it's probably not there.

Marginal Effects

Another argument we often hear, especially in parapsychology, is that a large number of marginal effects can add up to positive evidence for a new phenomenon, even when no single effect is individually significant. This is a common misconception. Marginal effects are often seen as a prelude to definitive experiments, but they never stand on their own. Sometimes they are confirmed; more often than not, they go away. I can think of no phenomenon in physics that was discovered without a definitive experiment to demonstrate a significant effect that could stand on its own two feet. Replication then follows as the final confirmation of the effect.

Recently in physics there have been published reports of magnetic monopoles, fractionally charged particles, and a fifth force, none of which have been independently confirmed. Perhaps one or more of these phenomena will eventually be validated. But our experience with most marginal effects is that they eventually fade away.

The Channelers

So no evidence for ESP is found in either my simple experiment or the many thousands of other experiments that have been conducted over the years. Still, belief in special powers of the mind persists. Individuals who claim such special powers continue to convince much of the public that they can read minds, perform mental miracles, or communicate with worlds beyond the senses.

We have seen how the mediums of the late nineteenth and early twentieth centuries were modern secular reincarnations of the religious mystics of previous centuries. By postwar years, these mediums were largely replaced by psychics who specialized more in personal readings and advice, less in seances. While people are still impressed by the mental tricks of psychic readings, the special physical effects of the seance pale by comparison with a George Lukas or Steven Spielberg film. The psychic fad now appears to have peaked and been replaced by what to me is little more than a New Age, space age, rerun of the old spiritualist medium. This is the "channeler."

Channelers go into trances similar to mystics or mediums. But instead of talking to the recently departed loved ones of their grieving clients, channelers often claim to be able to make psychic contact with supernatural entities thousands of years in the past or future, or thousands of light years away in space. These then provide advice to the channelers' clients. The channeler, like the medium, acts as an agent for some distant superentity, who usually speaks excellent English, though often with an exotic accent and in a strange-sounding voice reminiscent of the Devil in the film *The Exorcist.*

For example, Joan Davies's source is a 10,000-year-old spirit who gives such profound messages to her clients as "all things happen for a reason." Penny Torres, a Los Angeles housewife, channels the words of a 2,000-year-old-man named Mufu who is in contact with extraterrestrials living in colonies under the earth. Nevelle Rowe channels messages from dolphins off the California coast. Taryn Krive channels "Baby Bell," a child from the lost continent of Atlantis (Kaplan 1988). Greta Woodrew is in psychic communication with the "Ogata Group," consisting of thousands of advanced civilizations spread throughout the galaxy.

Although it has just recently become a craze, channeling in this modern form began in 1963 when a New York woman named Jane Roberts claimed contact with a spirit called "Seth." Through Roberts, Seth purportedly dictated a series of books on philosophy. Roberts also produced a book on art criticism said to have actually been dictated by the spirit of the artist Paul Cézanne.

The most successful channeler of recent years is a woman named J. Z. Knight, whose source is a being called "Ramtha." Ramtha speaks through Knight's mouth in an odd-sounding male voice. Knight, who was greatly boosted by Shirley MacLaine, grosses as much as $200,000 in one night.

However, Knight has been somewhat tarnished by revelations on the ABC network TV show "20/20" that as a child she would speak in prayer meetings in a male voice, purported to be that of a demon named Demias. It was also revealed that Knight can jump in and out of the Ramtha character without going into the elaborate trance used in her public performances (Kaplan 1988).

Linguist Sarah Thomason has performed a linguistic analysis on tapes of eleven channelers in action (see *Psychology Today,* October 1989, p. 64). She found that all had inconsistent and implausible accents or dialects, inappropriate for the time and place from which the sources were supposedly communicating. For example, channeler Marjorie Turcott's entity called Matthew is supposedly a sixteenth-century Scotsman. Yet he pronounces words like "neighbor" in the current English way, with a silent "gh," something not even a modern-day Scotsman would do—much less one who lived 400 years ago. Matthew also uses terms, like "rapscallion," that did not appear until much later.

Penny Torres's Mufu channels from the first century with a British accent that can go no further back than 800 C.E. Prior to that, English sounded closer to German than modern-day English. New York channeler Julie Winter's Mitka switches between American and British accents, saying "very" one time and "veddy" another.

Still, the believers believe. In what is labeled a "critical review" on channeling in *New Age Journal,* Jon Klimo concludes that "channeled material cannot be explained" as normal fantasies or "perceptions and memories derived from interactions with external reality," and that "the ego's interface with external reality somehow includes extrasensory or nonphysical means" (Klimo 1987).

Klimo claims that channelers obtain information that cannot be simply the product of sensory input or the channeler's imagination. I have earlier argued that history's mystics never successfully demonstrated that their revelations provided new truths about the universe. Rather, they seemed to produce images that ultimately bore little resemblance to the images eventually obtained by scientific instruments.

Now, perhaps the human mind has evolved to the point where the new mystics have powers never before realized. If that's the case, then we should have no trouble confirming it. Channeling fortunately lends itself to precise empirical tests. Although the skeptic has no responsibility for finding evidence to support paranormal claims, the burden of proof resting with the proponents of these claims, I would still be happy to provide the claimants some help in this respect. I will show a way that they can prove their claims beyond a shadow of doubt.

Challenging Channelers

A number of questions of fundamental and cosmic significance remain currently unanswered in particle physics and cosmology. The answers must surely be known to the entities from advanced civilizations now channeling to earthlings in New York, California, and Windward Oahu. I will list these questions here, and invite channelers to pose them to their contacts. If a reasonable number of the answers turn out upon future investigations to have been correct, then we might begin giving the phenomenon some careful attention.

The most important unanswered question with respect to the Standard Model concerns an important element of the theory that has not yet been observed: the Higgs boson (or bosons, if there are more than one). Thus, the first question I would ask a channeler's contact to provide is:

> 1. *What are the masses of the Higgs bosons and how many types of them are there?*

Going beyond the Standard Model, we would like to know how to unify all the forces of nature. So, please ask:

2. *Is there a single unifying symmetry group for all the forces? If so, what is it?*

This may seem like a question that relies on particular mathematical ideas developed on earth, but since channelers' sources speak English, they surely must know earthling mathematics.

We particle physicists accept the fact that our current picture of quarks and leptons as the elementary contituents of matter, highly successful though it is, is undoubtedly provisional. Thus, a very basic question that the supernatural entities can be a big help with is:

3. *What are the fundamental objects of the universe? Are they particles? Strings? Membranes?*

Going beyond matter to the nature of space and time,

4. *How many dimensions has the universe, including those which may be curled up on unobservable scales?*

And,

5. *Is the universe "open" or "closed"?*

Of course the superentity has a fifty-fifty shot at this one, but we would still like to know.

Another important piece of cosmic information we need is the precise rate at which the universe is expanding, which is currently uncertain by about a factor of two. So,

6. *What is the value of the "Hubble parameter"?*

As I have said, the best candidate for a possible anomaly lies in the nature of the dark matter of the universe that we believe is there and yet have not been able to detect; so,

7. *What is the nature of the dark matter of the universe?*

Back to particles, the answers to a few questions can be checked in the near future, when larger accelerators go into operation:

8. *What is the mass of the neutrinos, top quark, and "supersymmetric particles" (if these last exist)?*

So far we have found three generations of quarks and leptons and would like to know:

> 9. *How many generations of quarks and leptons exist?*

And,

> 10. *What is the mean lifetime of the proton?*

Finally, a long-standing puzzle is the fact that fewer neutrinos are observed from the sun than expected; so,

> 11. *What is the answer to the solar neutrino puzzle?*

Anyway, you get the idea. In fact, this might be used as a party game. Have the guests think of questions to ask channelers that would provide good tests of their paranormal claims.

If any channeler or psychic predicts even three or four right answers, I will become his or her number one believer. I may even obtain fame and fortune for having suggested the definitive series of tests that finally demonstrated the existence of a world beyond the senses. So I wish the channelers the best of luck!

References

Arguelles, José. 1987. *The Mayan Factor: The Path Beyond Technology.* Santa Fe, N.M.: Bear & Company.

Basil, Robert. 1988. *Not Necessarily the New Age.* Buffalo, N.Y.: Prometheus Books.

Dass, Ram. 1973. *The Only Dance There Is.* New York: Anchor Press.

———. 1977. *Grist for the Mill.* Santa Cruz, Calif.: Unity Press.

———. 1978. *Journey of Awakening.* New York: Bantam Books.

———. 1979. *Miracle of Love.* New York: Bantam Books.

Ferguson, Marilyn. 1980. *The Aquarian Conspiracy: Personal and Social Transformation in the 1980s.* Los Angeles: J. P. Tarcher.

Gardner, Martin. 1988. *The New Age: Notes of a Fringe-Watcher.* Buffalo, N.Y.: Prometheus Books.

Gordon, Henry. 1988. *Channeling into the New Age.* Buffalo, N.Y.: Prometheus Books.

Kaplan, Steve. October 1988. "The Channeling Craze." *The World & I.* P. 286.

Klimo, Jon. November-December 1987. *New Age Journal.*

Lasch, Christopher. October 1987. "Soul of a New Age." *Omni* 10:82.

MacLaine, Shirley. 1983. *Out on a Limb.* New York: Bantam Books.

———. 1985. *Dancing in the Light.* New York: Bantam Books.

———. 1987. *It's All in the Playing.* New York: Bantam Books.

10.

The Spooks of Quantum Mechanics

I cannot seriously believe in [the quantum theory] because it cannot be reconciled with the idea that physics should represent a reality in time and space, free from spooky actions at a distance.

Albert Einstein

Whistling in the Graveyard

The "spooky actions at a distance" that so disturbed Einstein about quantum mechanics really exist. They have, in fact, been confirmed in a remarkable series of experiments. This result has given comfort to those who seek a scientific basis for spooky things like telepathy and faster-than-light communication, as well as the holistic notion that everything that happens in the universe is simultaneously connected to everything else.

Do paranormalists have a right to be comforted, or are they just whistling in the graveyard? I'm afraid that they would have been better off if things had turned out not quite so spooky. As we will see, the conventional interpretation of quantum mechanics, disputed by Einstein but triumphantly confirmed by experiment, provides no mechanism for psychic phenomena or simultaneous connections between events. On the contrary, the foundational principles of twentieth-century physics, relativity and quantum mechanics, have been confirmed by countless empirical tests since they were first proposed early in the century. They have stood up under every challenge scientists or pseudoscientists have been able to mount.

Light is a Wave

Einstein's unhappiness with quantum mechanics is legendary, though his photon theory of light contributed mightily to the quantum revolution. In the seventeenth century, Isaac Newton proposed that light was "corporeal" in nature, that is, composed of material bodies. By the nineteenth century, however, the wave properties of light were well established, and Newton's corpuscular theory of light was largely dismissed and replaced by the wave theory of light first proposed by his contemporary, Christiaan Huygens (1629–1695).

The word "particle" implies the notion of "locality." Particles are discrete bits of matter confined to small regions of space. They carry energy, momentum, and other quantities, such as electric charge, from one point in space to another by physically moving between the points.

Another means for the transfer of energy and momentum from one point in space to another is the familiar one of wave motion in a continuous medium. The vibration of the medium can result in mechanical waves, such as ocean or sound waves, that transfer energy without actual movement of any individual piece of matter. The medium as a whole, not just a small part of it, participates in the process.

The air, water, and earth of common experience appear continuous to our eyes. However, by the beginning of this century evidence had accumulated to establish beyond a shadow of doubt the ancient idea that everything we normally classify as matter is composed of many billions of discrete particles: atoms. Still, even as recently as 1900, light seemed to be a different form of reality, a part of the universe separate and distinct from matter. Light was thought to be the vibration of an invisible medium that pervaded all of space: the "aether."

Light is Particles

However, at the turn of the century, a few curious observations of no obvious practical import stubbornly remained unexplained by the wave theory of light. First, the spectra of light that radiated from hot solids and gases showed individual narrow lines of specific wavelengths, instead of the continuous band predicted by the wave theory. Second, the photoelectric effect, in which an electric current is triggered when light above a threshold frequency impinges on certain materials, could not be reconciled with the description of light as a wave.

Building on Max Planck's five-year-old idea that light occurs in discrete bundles, or "quanta," in 1905 Einstein resurrected Newton's corpuscular theory of light in an updated form, and proposed that light was composed of localized particles he called "photons." Today it is unquestioned that light consists of discrete particles. The fact is confirmed daily in thousands of laboratories. Modern photomultiplier tubes and charge-coupled-devices (CCDs) easily register individual photons.

The Wave-Particle Duality

Empirically, light exhibits the discrete and local properties associated with particles. But apparent wave characteristics of light, such as its ability to pass simultaneously through several openings separated in space, are also confirmed.

This type of schizophrenic behavior is not confined to photons alone. Electrons, neutrons, and other entities that normally appear as localized particles also can't seem to decide whether they are waves or particles. It all depends on what you try to measure. If you look for localized electrons, neutrons, or photons, you find them. If, on the other hand, you set up an experiment designed to measure wave properties, you find these too. We look at the world through colored glasses, and so it should not surprise us that the world appears a different color when we change to another pair.

This wave-particle duality, exhibited by normal matter as well as light, was first recognized by Louis de Broglie in 1924 and set the stage for subsequent independent developments of quantum mechanics by Werner Heisenberg and Erwin Schrodinger. This duality is precisely what gives quantum mechanics its spooky nature.

One should not confuse the wave-particle duality with the matter-spirit or mind-matter dualities, or with some vague notion of "mind waves" of "pure spiritual energy." As I have already noted, matter and energy are the same stuff.

The Definition of Physical Quantities

In his 1905 special theory of relativity, Einstein emphasized a feature of scientific method that was to become a key to the development of quantum mechanics: physical quantities are defined by the way they are measured. This includes those concepts that most still regard as possessing an inherent reality, such as time and space.

Einstein reminded us that a time interval is what is measured on a clock and a space interval is what is measured with a meter stick, and then showed us that these depend on the relative motion of the source and observer, in violation of common sense. This was consistent with a new school of philosophy that had been developing at about the same time, called logical positivism.

Positivism held that the only reality was empirical observation, and that metaphysical notions were nonsense, since they dealt with concepts that had no empirical content and so could not be tested observationally. The basic idea, which seems emminently reasonable to me, is that untestable propositions about entities that have no effect on the observable world are meaningless. Today logical positivism is out of fashion, but any new philosophical system must still confront the question of how to make meaningful statements about the universe that are not directly related to empirical facts.

Although Einstein had provided the greatest impetus to the positivist view of nature, what quantum mechanics did with space, time, and all the other quantities of physics went far beyond the great physicist's surprisingly conservative tastes. In the years following his immortal 1905 papers on relativity, Einstein backed off considerably from his original positivist views, insisting that physical properties must have an intrinsic reality beyond their mere measurement.

This change of heart was the result of his development of the general theory of relativity, first appearing in 1916, in which space and time seem to have intrinsic holistic reality. As he says, "My departure from positivism came only when I worked out the general theory of relativity" (Einstein 1948). As a result, Einstein ultimately wound up on the losing side of the debate on the nature of quantum mechanics and what it tells us and doesn't tell us about the universe.

The Copenhagen Interpretation

Niels Bohr, Max Born, and the other developers of what has become the most widely accepted view of quantum mechanics, called the Copenhagen Interpretation, held that an object does not even possess certain properties until those properties are measured. When we observe anything in the universe, we must bounce photons or some other particles off it. That is, we must interact with the object being observed, disturbing it in some way. This presents no problem for large objects, like the moon and most bodies in our everyday experience that hardly recoil under the action of the photons from the sun or artificial lights. However, at the atomic and subatomic level, photons can wreak havoc with the system being observed.

For example, the position of an electron inside an atom cannot be measured by bombarding it with photons—without destroying the atom. Consequently, the Copenhagen Interpretation holds that that an electron actually has no specific position while it remains inside the atom. In other words, since the electron's position is unmeasurable, it is therefore meaningless. This is not just cocktail party chatter. The attraction between nearby atoms, which holds them together into the molecules that make up each of us and the things we see around us, is at least partially the result of the fact that an electron cannot be located in one atom or the other.

Extending this idea to all other physical quantities, we conclude that they become real only upon being measured. Now it may sound as if I am adopting ancient Hindu idealism, with everything in our heads after all. As I have noted, this view is also a tenet of the New Age: Reality is what you want it to be. The fact that reality rarely is what you want it to be is the best evidence that a world beyond our heads does indeed exist.

However, the reality I am concerned with here is specifically that of physical quantities: the variables we use in the descriptions of our observations in physics, such as space, time, mass, and temperature. According to the Copenhagen Interpretation, these are human inventions and are defined according to the way they are measured.

This is not to say that the world is all in our heads. The world has objects we can label as reality. These objects have properties we identify and describe in terms of physical variables that are determined by measurement. As I will show, however, observations force upon us the conclusion that the variables we measure have no meaning beyond their actual measurement.

Einstein rejected the Copenhagen Interpretation, saying that an object must have an objective reality of its own independent of the act of measurement and, by inference, the actions of the person doing the measuring. The moon, he remarked, exists even when no one is looking at it.

But as we will see, quantum mechanics does not reject the existence of external objective reality. The last thing any scientist should want to do is return us to the idealist notion that everything is in our heads. Conventional quantum mechanics simply says that the *physical quantities* we use to describe reality are meaningless except as they result from a carefully prescribed, but nonetheless arbitrary, measurement procedure. This is the view that was questioned by Einstein in 1935, but is now confirmed by experiment.

The Overthrow of the Clockwork Universe

Another important element of the Copenhagen Interpretation of quantum mechanics is that the equations of physics do not allow you to predict the movement of an object from one place to another with complete certainty—only the *probability* of this movement. This probability is expressed in terms of a purely mathematical quantity called the "wavefunction"—

so-called because it has the same mathematical characteristics as the quantities that are used in wave theory to describe mechanical waves.

The idea that the motion of individual bodies is at least partially unpredictable, and thus at least partially undetermined by what happened before, contradicted the assertion of Newtonian mechanics that all motion is in principle completely predictable, provided you have sufficient information to make the calculation.

In Newtonian mechanics, the universe is a vast mechanism, a clockwork, with the behavior of every material object totally determined by the forces acting on it. The Copenhagen Interpretation of quantum mechanics says that this is not so, that an intrinsic randomness and indeterminacy exists in nature. The equations of quantum mechanics only allow you to predict the average behavior of similarly prepared systems, not the exact behavior of the individual systems themselves.

Hidden Variables

Shortly after the development of quantum mechanics, the discoverer of the wave-particle duality, Louis de Broglie, had proposed that particles were guided by pilot waves and that these were responsible for the observed wave-like behavior of electrons and other particles (De Broglie 1930). He associated the quantum mechanical wavefunctions with his pilot waves. But Heisenberg and others argued that this was unlikely, since, in quantum mechanics, a single wavefunction is used to describe systems of many particles.

In his excellent 1951 textbook *Quantum Theory,* David Bohm developed the De Broglie idea further, proposing that some form of hidden variables may exist that provide for a more complete description of nature than conventional quantum mechanics—more complete in the sense that the statistical nature of quantum mechanics is replaced by underlying deterministic principles. The hidden variables would be analogous to the forces and potentials of classical physics that are responsible for the motion of bodies (Bohm 1951).

In the intervening years, Bohm has led a small group of investigators who have attempted to find a deterministic alternative to conventional quantum mechanics using the framework of hidden variables (see, for example, Bohm, Hiley, and Kaloyerou 1987). In most hidden variables theories, the motion of particles is determined in a Newtonian way, and the apparent randomness we observe is the result of our not having adequate apparatus to detect the hidden variables directly. However, we will soon see that the commonsense-type hidden variables, based most directly on ideas carried over from classical physics, are now ruled out by experiment.

The EPR Paradox

In 1935, Einstein and two junior colleagues, Boris Podolsky and Nathan Rosen (I will refer to this team as EPR), wrote a paper arguing that quantum mechanics had to be "incomplete" in its description of reality. While they did not dispute that quantum mechanics was correct as far as it went, they asserted that some underlying reality not described by the conventional theory must exist or else one must discard some rather basic notions of reality.

EPR argued that certain quantum systems, such as those composed of two particles, can be prepared in such a way that the result of a measurement of one particle fixes the result of a measurement of the second— before that measurement is performed. This can happen after the particles are so separated in space that communication between them would require a signal traveling faster than light. EPR concluded that either the particles intrinsically possessed certain properties before they were measured—in contrast to the quantum idea that the act of measurement brings that property into existence—or a nonlocal "spooky action at a distance" force was in effect. This has become known as the "EPR paradox" (Einstein, Podolsky, and Rosen 1935).

Reality

Two concepts play a crucial role in understanding the EPR effect and its consequences: (1) *reality* and (2) *locality*. Reality is another one of those words in common use that becomes harder to define the harder we think about trying to do so. So it is interesting to note how Einstein and his colleagues defined reality for the purposes of their discussion: "If, without in any way disturbing a system, we can predict with certainty (i.e., with probability equal to unity) the value of a physical quantity, then there exists an element of physical reality corresponding to this physical quantity."

EPR admitted that this is neither a comprehensive definition, nor does it exhaust all the other possible ways of recognizing physical reality, but it provided them with a clear empirical test that contains within it a reasonable criterion for labeling something as real.

The concrete objects of our everyday experience contain this property of reality. If we look away from a tree, we can predict with almost absolute certainty that it will still be there when we look in that direction a few moments later. For moving objects, we can use the laws of mechanics, or our own experience, to predict their motion with high certainty. In my youth, as an outfielder, I could turn my back to the plate at the crack of the bat, run to where I had calculated the ball was going, and turn around and catch the ball as it came down. Certainly that baseball was a real object to me, as I felt the sting of it hitting the pocket of my glove.

Locality

To see what is so spooky about the EPR paradox, we also have to understand precisely what we mean when we say that two events are local. Suppose I throw a ball from here and you can catch it over there. To each of us, these two events—my tossing the ball and your catching it—occur at different places. As a result, you might say these events are nonlocal, that the ball performed an "action at a distance."

However, consider the point of view of a third observer, who is moving along with the ball at exactly the same speed. She will experience both events to occur at the same place in her personal frame of reference.

Now, one of the basic tenets of relativity is that our descriptions of nature cannot depend on the choice of reference frame. When we can find a reference frame, any reference frame, in which two events occur at the same place, we are forced to conclude that the two events are *local*. In that reference frame, no action at a distance took place. Moreover, since our principles of physics should be independent of any particular reference frame, we must regard the event as local in all reference frames.

Before Einstein and relativity, it was thought that no limit existed on the speed of bodies. Thus, whatever the separation of two events in space and time, it was always possible to find a reference frame in which the events were local. Even events occuring at the same time kilometers apart would be local to an observer moving at infinite speed.

However, Einstein's special theory of relativity requires that no object can travel faster than the speed of light. If this is correct, then there must exist certain events that cannot be made local in any reference frame. These events need only be separated by sufficiently great distances and small enough times that an observer would have to move faster than the speed of light to see each at the same position.

For example, events on the earth and the moon might occur within a time interval of one second as measured by a clock on earth. However, since the moon is 380,000 kilometers away, and light only travels at 300,000 kilometers per second, an observer would have to travel at greater than the speed of light to experience both events in the same position in his reference frame. In relativity jargon, such events are said to be "outside the light cone."

Inside the Light Cone

In the conventional physics view, events outside the light cone cannot in any way be causally connected with one another, for to be so would require signal transmission at faster than the speed of light. The "spooky actions at a distance" that concerned Einstein were causal connections between events that lay outside the light cone.

If the theory of relativity is correct, then there can exist phenomena in the universe that are independent of one another—namely, those events that are outside the light cone. A special set of events outside the light cone, very important for the types of ideas we have been considering in this book, are those occurring simultaneously at different places. Simultaneous events that are separated in space in any reference frame cannot be causally connected with one another. That is, they cannot affect one another in any way.

Thus, the holistic notion of a simultaneous connection between all events separated in space is inconsistent with Einstein locality. To be consistent, time must be allowed for signals to travel between events. Connections can exist between events that are separated in a given reference frame. Indeed, such connections exist. But these connections are local only if a reference frame can be found in which the events occur at the same place.

Holism Violates Einstein's Relativity

This is a point that hardly anyone, including most scientists, recognize: *holism violates Einstein's theory of relativity.* If simultaneous holistic connections between separated events exist, then either the whole foundation of twentieth-century physics must be destroyed or these connections must be supernatural.

Natural events, consistent with known physics, all lie within the light cone. If the universe began in a tiny region of space, all of it remains inside the light cone whose apex is the beginning of space and time. Separated events inside the universe can be, and of course are, connected, but those connections are all local in the Einsteinian sense.

This is precisely why I object so vehemently to the vague holistic principles espoused by so many who are associated with paranormal thinking. These principles sound good when sitting around in a coffeehouse, but they send arrows to the very heart of physics. Perhaps those arrows will someday be justified. But that justification will need to have a foundation in empirical fact. Put another way, the very success of conventional physics, including Einsteinian locality, in describing all currently known empirical data makes the existence of simultaneous or superluminal connections very unlikely.

Bohm's EPR Experiment

In his 1951 book, David Bohm intuitively suggested that "no theory of mechanically determined hidden variables can lead to *all* of the results of quantum theory." Moreover, Bohm identified a testable difference between the two points of view. He proposed an EPR-type experiment involving electron spins that would provide the conclusive test (Bohm 1951). However, it took another thirteen years before anyone could prove it.

Spin

Bohm recognized that the key was "spin": the intrinsic angular momentum of a particle. Spin has components along each of the three axes of space: x, y, and z. According to quantum rules, however, spin has a very strange property: only one of its three components can be measured at a given time.

When one spin component is measured, the other two components can have no meaningful value. Measuring another component upsets the measurement of the first, as can be demonstrated by going back and trying to measure the first component again; you often get a different value. It's as if you measured the height of a woman to be 5 feet 4 inches, then her waistline, and then went back and measured her height again and found it to be 4 feet 5 inches. Crazy as it sounds, this is exactly what happens, in the laboratory, with the components of spin.

Bohm's Thought Experiment

Bohm suggested the following thought experiment: suppose you start out with two electrons in a certain state in which the total spin is zero, the so-called singlet state. Then let the electrons go off in opposite directions. After they have separated by some distance, measure the spin component of one electron along a particular axis, say the z-axis (see Figure 10.1). As long as no external interactions able to impart or remove angular momentum take place, total angular momentum is conserved, and the total spin of the two-particle system must remain zero, even as the particles become separated.

As a result, the measurement of the z-component of the spin of one electron immediately fixes the z-component of the spin of the other electron. The measurement of the first allows you to predict with absolute certainty the outcome of a measurement of the other. And so, in the EPR sense, the z-component of spin corresponds to "an element of physical reality."

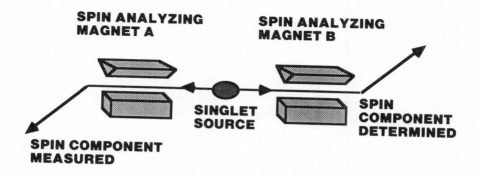

Figure 10.1. The EPR Paradox experiment proposed by David Bohm. Electrons from an initial singlet (total spin zero) state are analyzed by magnets that are oriented to deflect the particles one way when they are spinning along a particular axis, and the opposite way when they are spinning opposite. Once the spin component at one end is measured, the component at the other end is determined. In conventional quantum mechanics, spin components do not exist until they are measured.

But what about the x- and y-components of the electron's spin? In classical physics, all three components are intrinsically real, as are the three components of the angular momentum of the earth. The electron simply carries along these properties as it moves from the source to the detector, just as it carries its mass and electric charge. However, conventional quantum mechanics says that only one spin component can be intrinsically real at any given time, specifically, the last one measured. That particular measurement axis, totally arbitrary, is called the axis of quantization.

In the case of the Bohm experiment, the experimenter at the end of one of the electron beams can choose any axis he wants. When that spin component of one electron is then measured, the corresponding spin component of the other electron instantaneously becomes determined! An observer at the end of that beam line who then proceeds to measure the spin along the selected axis will, with 100 percent certainty, obtain the value specified. This could happen at a sufficiently large distance and small time interval to be outside the light cone. Somehow the information that the first electron's spin has been measured is transmitted instantaneously to the second. Thus, Bohm's experiment nicely demonstrates Einstein's "spooky action at a distance."

Real, Local, Hidden Variables

The idea of hidden variables provides several possible interpretations of the EPR paradox, but let me focus on only one that is consistent with both Einsteinian locality and EPR reality, and our own intuitions carried over from classical physics. In this interpretation, the hidden variables are both real and local.

Real, local, hidden variables are variables that: (1) connect events in space only within the light cone, and (2) have intrinsic reality that we can take to be defined in the EPR sense. They correspond to many of the types of physical quantities that are familiar from classical physics, for example, mass, electric charge, energy, and angular momentum.

Applied to the Bohm thought experiment, the hidden variables could be the three spin components themselves. Since they cannot be simultaneously measured, an experimentally demonstrated fact, at least two must be "hidden." Conventional quantum mechanics, on the other hand, says not that they are hidden, but rather that they are *not real,* because only measurement can define whether a variable is real.

The hidden variable interpretation disagrees, saying that reality goes beyond measurement. Which interpretation do you prefer? The spooky one of Bohr and Born or the sensible one of Einstein, De Broglie, and Bohm? How can we decide? The same way we decide anything in science—by experiment.

Bell's Theorem

In 1964, John S. Bell analyzed Bohm's thought experiment and was able to prove that, with modifications, it provided a practical means for distinguishing conventional quantum mechanics from real, local, hidden variables theories. Basically, Bell proved mathematically that the quantitative correlation between measured spin components of the two electrons from a singlet source that is computed from quantum mechanics is greater than would be possible in any theory of real, local, hidden variables (Bell 1964, Clauser and Shimony 1978, d'Espagnat 1979, Redhead 1987).

The term "quantitative correlation" applies to a quantity that is a measure of how closely the value of one variable depends on the value of another. For example, one might compare the average age difference of married couples with that for all pairs of persons in the entire community. The first would be smaller than the second, indicating a stronger correlation between the ages of married couples than for all people in general.

Similarly, the average value of the product of the electron spin components along the x and y axes constitutes a measure of their correlation.

Since the components are equally likely to be plus and minus in a case where they exist in a random mixture, the average would be zero in this case. A nonzero measurement would provide evidence for a nonrandom, or correlated, mixture.

According to Bell's theorem, if the Einstein-Podolsky-Rosen-Bohm-Bell experiment were performed and the results were consistent with the predictions of the statistical interpretation of quantum mechanics, then this particular class of hidden variables, local and real, would be ruled out.

This was a rather remarkable result. The debate between the conventional and hidden-variables schools of quantum mechanics might strike you as the sort of question that theologians used to argue over, like the number of angels that can dance on the head of a pin. Yet it can actually be objectively settled in the physics laboratory. Once again, the power and beauty of the scientific method as a means for the determination of truth about the universe becomes evident.

Bell determined the correlations that would occur in the circumstance that all three spin components were intrinsically real in the classical sense discussed earlier, where the three components can have values even when unmeasured. In certain cases, this correlation proved to be lower than what one would calculate using conventional quantum theory. Bell's result was expressed as a mathematical inequality.

Conventional quantum mechanics violated Bell's inequality, giving correlations over a distance that were fundamentally impossible with localized, realistic variables. Since the theorem is quite specific mathematically, Bell showed that the fundamental local and real character of physical variables, such as spin, can be conclusively tested by experiment.

A Simple Spooky Experiment

Before discussing the actual experimental tests of Bell's inequality, it will help to set the stage to show that the same kinds of spooky effects attributed to quantum mechanics were already observed for the phenomenon of light, long before anyone thought about quantum mechanics. This will make it easier to understand why the results of the tests of Bell's inequality are really not as strange as they appear at first glance.

We can see spooky effects in a simple light polarization experiment that may already be familiar to the reader. To perform this experiment, you only need a few cents worth of polarizing plastic, obtainable from any optical supply outlet. Designate three pieces of the polarizer A, B, and C. Set C aside and overlay the other two. Holding them up to a lamp, rotate one with respect to the other until they block off most of the light passing through (see Figure 10.2). Now, keeping this relative

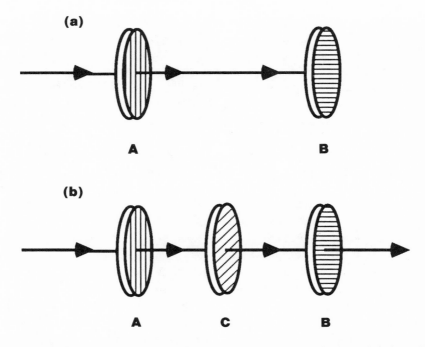

Figure 10.2. The crossed polarizers A and B in (a) block light from passing through. Inserting polarizer C in between with polarization direction at an angle, as in (b), causes the light to reappear.

orientation fixed, carefully insert piece C between the pieces A and B. Then, rotating the central piece without disturbing the other two, you will see light magically reappear, passing through the three pieces with an intensity that depends on the angle of rotation of C with respect to the other two.

In other words, by inserting a piece of plastic between two other pieces that together are opaque to light, we have made them once again transparent. This is contrary to common sense. Intuitively we might expect the insertion of additional material in a light path would reduce the amount of light transmitted rather than allow light that was previously blocked to pass through.

When this experiment is performed in a junior high school science class, the teacher usually explains how sheet A polarizes the light along one axis, so that when sheet B is rotated so its polarization axis is at right angles to A, no light passes through. But, inserting sheet C in between A and B rotates the plane of polarization, so that a component of polarized light along the axis of the C sheet is created, and this component is able to pass all the way through.

Note that no quantum mechanics is used in this explanation, and indeed the effect was observed and explained in the last century. However, some of the spooky stuff of quantum mechanics can be understood in the same nonspooky nineteenth-century way: the wavefunctions of quantum mechanics are analogous to the classical electric and magnetic fields that describe the wave nature of light. So they can be no more spooky than those more familiar fields.

In the classical wave theory of light, an unpolarized light beam is composed of many uncorrelated (incoherent) waves. The polarizer allows only those waves to pass through whose field vectors point in the right direction. The initial light beam was a composite entity, originally containing both the waves absorbed and those that were transmitted, so nothing spooky was involved.

Circular Polarization

When the electromagnetic field vectors of a light beam remain in a plane fixed at a particular orientation with respect to the beam, we say that the light is linearly polarized. Another form of polarized light, circular polarization, results when the field vectors rotate around the direction of propagation, going once around for every cycle of the wave (see Figure 10.3 b). Looking toward an oncoming beam of light, the field vectors rotate clockwise for right circular polarization and counter-clockwise for left.

Circular polarized radio waves are produced by a loop (magnetic dipole) antenna. Linear polarization is simpler in the classical electromagnetic wave picture of light, being described by field vectors that remain fixed in orien-

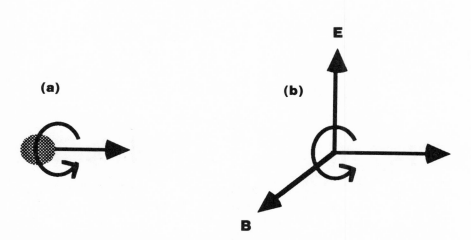

(a)   **(b)**

Figure 10.3. The right-handed photon shown in (a) is spinning around an axis along its direction of motion. This corresponds to a left-circularly polarized electromagnetic wave in which the field vectors E and B rotate counterclockwise around the direction of propagation of the wave, as in (b).

tation as the wave propagates. They are produced by a straight (electric dipole) antenna.

The two forms of polarization are related. Any orientation of linear polarization can be described as a combination of the two circular polarization states. The opposite rotations of the field vectors cancel each other out (though not their oscillations), so that the net field vector oscillates back and forth in a plane at a fixed angle with respect to the beam. Any value of this angle is possible, so an infinite number of combinations of the two circular states can occur.

We now modify our experiment slightly. Let sheets A and C each be circular-polarizing rather than linear-polarizing, while B remains linear-polarizing. Polarizer A will then only transmit, say, left-circular light. So if sheet C lets through only right-circular light, no net light transmission will occur. Next, when we place sheet B (still a linear polarizer) in between A and C, it will pick out the linear component of the circularly polarized light (see Figure 10.4). Remember, each type of polarization is a superposition of the other types. The light intensity will be reduced corresponding to that component transmitted. The light transmitted through B will now be linearly polarized and thus contain both right and left circularly polarized components. The left circular part will then pass through polarizer C.

Polarizing Photons

The completely classical—meaning not quantum mechanical—explanation of the polarized light experiments presents no problem as long as we use comparatively insensitive light sensors, like the human eye. But suppose

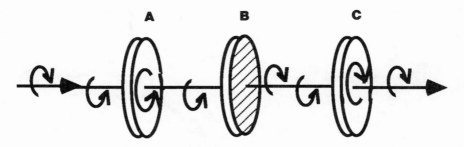

Figure 10.4. Unpolarized light becomes left circular polarized in passing through A. After passing through linear polarizer B, a right circular component is re-introduced which then passes through C.

we use photomultiplier tubes or charge-coupled devices. Then we must confront the matter of a single photon, which would be easily detectable by these instruments.

In the photon theory of light, the polarization of the electromagnetic waves is replaced by the spin of the photon. Recall that only one component of spin can be measured at a time, and the axis along which we choose to make that measurement is called the axis of quantization.

We say a photon is right-handed when the axis of quantization is along its direction of motion. Suppose we oversimplify a bit and think of the photon as a spinning sphere. If you point the thumb of your right hand in the direction of motion of a right-handed photon, your fingers will point in the direction of rotation (see Figure 10.3 a, p. 227). Left-handed spin is in the opposite direction. In both unpolarized and linearly polarized beams of many photons, we have half of each.

Unfortunately, we now have a minor collision of definitions: a beam of right-circularly polarized light contains only left-handed photons, while a beam of left-circularly polarized light contains only right-handed photons. This may be confusing, but it is not important.

In our second polarizer experiment (Figure 10.4), the initially un-polarized beam is composed of a stream of individual photon particles, half spinning in the direction of motion and half spinning in the opposite direction. The first polarizer selects only those that are right-handed, spinning in the direction of motion. How is it possible then that a left-handed photon, spinning opposite the direction of motion, can emerge from polarizer C, when all the photons out of A were right-handed?

Conventional quantum mechanics offers the following explanation: the ensemble of photons at any point in the beam is described by a wavefunction, a quantized version of the classical electromagnetic field vector, with left and right components whose absolute squares give the relative number of photons with a particular polarization. Only right-handed photons pass though A, so the wavefunction becomes right-handed by virtue of the act of measurement, just as in classical theory the field vector is rotated in passing through the polarizer. All this means is that another right-handed polarizer will allow most of the remaining photons to pass through.

Now the right-handed wavefunction will contain a combination of all possible states of linear polarization. So, when the photons pass through polarizer B, the component of the wavefunction that is linear in that direction is selected. This makes the wavefunction linearly polarized, and as such it now contains a component of right-handed photons that can be measured by polarizer C.

In a beam of photons, the measured intensity of light transmitted will be proportional to the number of photons passing through, as given by the square of the wavefunction. Now here's the important point of all of this: the photon intensity is exactly equal to the classical light intensity, being given by the absolute square of the field vector. In the typical light beam, containing billions of photons, fluctuations are so small as to go unnoticed, and the intensity appears constant and precisely calculable from electromagnetic theory.

What happens to any individual photon, however, is not predictable in the conventional interpretation of quantum mechanics. Whether or not a particular photon actually makes it all the way through is then a matter of chance, undetermined by any law of physics, the outcome depending on the probabilities that the methods of quantum mechanics tell us how to calculate.

Experimental Tests of Bell's Inequality

Now let us return to Bell's theorem, and the experimental investigations of the EPR paradox that have provided a fascinating glimpse into the quantum world. At least a dozen experiments have been developed to perform the tests provided by Bell's theorem. The definitive series was performed by Alain Aspect and his collaborators at the Institute for Applied Optics of the University of Paris, Orsay, France (Aspect, Grangier, and Gerard 1982).

The Bohm-Bell proposed experiment involved electrons, but photons provide an equally powerful test. Fortunately, our previous discussion of spooky experiments with photons has set the stage for a quick grasp of these remarkable experiments. In the Orsay experiments, an atomic source emits pairs of photons in opposite directions. The wavelengths of the photons are in the visible region of the spectrum, that is, they correspond to visible light. The photons are produced by the decay of a singlet (spin zero) atomic system. Thus, when one photon is later measured to have a spin component along a particular axis, the other photon will have the opposite spin component, as with the electron experiment proposed by Bohm and analyzed by Bell.

Each beam passes through polarizers of variable orientation around the beam line, and then through photomultiplier tubes capable of counting individual photons. A different polarizer orientation is randomly chosen for each pair emitted from the source. In an advanced version of the experiment, Aspect and his collaborators arranged that the choice of polarizer orientation was made after the photons left the source. This provided an unambiguous test of Einsteinian locality, since in this case no signal could pass between the ends of the beams without traveling faster than the speed of light.

Using photomultiplier tubes as the basic light sensors, the experiment counted the number of photons obtained for each relative orientation of the polarizers at the ends of the two beams. From these counts, the experimenters could determine the amount of correlation between the settings at the end of one beam and the other.

Quantum Mechanics Emerges Triumphant

The results of the Aspect experiment showed greater correlations between polarizer orientations than were possible if all three components of the spin of each photon were intrinsically real, as defined by the EPR criterion. That is, Bell's inequality was found to be violated, empirically demonstrating that the spin components of photons cannot be both local and real. Furthermore, the quantitative correlation agreed precisely, within very small experimental errors, with that calculated using conventional quantum mechanics without the assumption of real spin components. Conventional quantum mechanics emerged triumphant.

But this also meant that we still had spooky action at a distance. Measurements at the end of one beam seem to determine measurements at the other, within time intervals so small that any signals between the two would have had to exceed the speed of light. Is this a possible mechanism for ESP and other psychic effects? Some have claimed so (Sarfati 1977, Herbert 1985, Targ and Putoff 1977).

Superluminal Communication

Let us see if an EPR device could be built to provide a means of superluminal signaling. Suppose we have a singlet source of pairs of photons or electrons that sends the particles of each pair in opposite directions. At the end of both beam lines we have detectors that can measure the spin components along axes at any angle in a plane perpendicular to the beam lines.

When the settings at both ends are the same, the spin component at one end is completely determined by the setting at the other. Thus, it seems that we should be able to use this fact to signal from one end of the beam to the other, instantaneously over great distances.

But it turns out we can't! In order to communicate information from one point to another, a series of bits encoding a nonrandom message must be transmitted. Leaving both detectors at the same angle provides for no information transfer. The receiver will simply find half of the particles with spin components in the selected direction, and half in the opposite direction.

To encode and decode messages, the detector angles of both sender and receiver must be changed. Now, the observer at the end of the receiving beam line cannot know ahead of time what angle to use—that information must be contained in the coded message he would receive. So the receiver would have to randomly change the detector angle and then try to extract the message from the number of received hits at each angle. One might be inclined to think that this would be possible, given the fact that the measurement of a photon or electron at one end determines the outcome of a measurement of the other member of a singlet pair. But it simply does not work out that way.

With each angle setting of one detector, a specific, calculable distribution of particles will be found for each setting of the other. In the case of photons, the relative numbers of photons observed at each setting will exactly trace the intensity patterns calculated from classical optics. It follows that the bit sequence formed by the individual photon counts at the receiving end will have no information content. As in the case of fixed predetermined settings, the bit pattern will be purely random.

Recall that experiments with polarized photons are like experiments with circular polarized light. If it was impossible to communicate superluminally with circular polarized light before the quantum theory, nothing about quantum theory has changed that fact. The wavefunctions are, for this purpose, equivalent to electromagnetic waves, and the photon distributions simply trace the identical intensity patterns we observe in classical optics. Again, no information is obtained in the observed distribution that we don't already know from the classical light intensity calculation.

Denying Reality

The results of the experimental violation of Bell's inequality are clear and unequivocal: nature cannot be described in terms of physical variables, hidden or visible, that are both local and real. Either Einsteinian locality or certain commonsense notions of reality must be discarded. There is no third alternative.

Most popular writers prefer to discard locality rather than reality (see, for example, Herbert 1985). They cannot understand why physicists are so unwilling to give up locality.

My answer is simple. We have no reason, indeed no right, to do so. Einsteinian locality is completely consistent with everything we know about the empirical world, from the nineteenth-century experiments that failed to find the aether to the tests of Bell's inequality. We have just seen that the apparent "spooky action at a distance" of quantum mechanics does not provide for superluminal transfer of signals, and so does not violate Einsteinian locality. The only violation of locality occurs within the imaginary framework of our own descriptions of the observations.

Abiding by Occam's razor, the more economical hypothesis is to deny the intrinsic reality of physical observables. Now, at first blush, this may not seem very economical. I appear to be making the most drastic of hypotheses. This guy is doing away with reality!

But actually I am being completely economical. I have, in fact, introduced no new hypotheses beyond those already implicit in relativity and quantum mechanics. Someday these may turn out to be wrong, but for now they are verified by decades of experimentation and practical application.

To toss out locality and demand reality, we make two unacceptable and unneccessary drastic hypotheses: that the two pillars of twentieth-century physics, relativity and quantum mechanics, as conventionally interpreted, are wrong. I would rather assume they are right, since they have been confirmed repeatedly, including by the experimental tests of Bell's theorem.

I am not in conflict with the commentators who say that Einstein was right in insisting on the existence of an underlying reality. I agree that the moon is there when no one is looking at it. I simply argue that the true reality of the universe is not necessarily composed of objects that possess attributes such as position and mass which we assign them in the process of doing physics. These variables, after all, are human inventions with no precisely definable meaning beyond their measurements as performed with specific apparatuses such as clocks and meter sticks.

Describing nature in terms of physical variables is like sketching or photographing an object. Isn't it rather foolish to equate images on a piece of paper with the real thing? Confusing an image with reality is a common characteristic of small children. We adults have supposedly outgrown their confusion.

We laugh at those ignorant and superstitious people who stick pins in dolls and think that this will harm the person represented. Yet even the most sophisticated physicists ascribe a kind of voodoo reality to their own mathematical images. Quantum mechanics is not voodoo, despite what books on the occult shelves of bookstores say.

No Support for the Paranormal

Paranormalists are dead wrong when they claim that relativity and quantum mechanics support their proposal of extrasensory channels in some kind of underlying aethereal reality. In fact, they had better start looking elsewhere for such support. Einstein's theory of relativity destroyed the aether and the instantaneous connection between events. All we know about science provides a picture of a universe of discrete objects, interacting with each other within the light cone.

Similarly, quantum mechanics cannot be used as a mechanism for supersensory phenomena. The experimental results of Aspect and his collaborators, and others, fully support the conventional statistical interpretation of quantum mechanics, with the wavefunction describing ensembles rather than individual particles.

While not all possible hidden variables theories are ruled out, the fact is that the distribution of photons is as predicted and thus incapable of carrying information. Even if some other form of hidden variables were someday to be found, they would not make possible superluminal communication by the mechanisms now proposed.

Quantum effects are certainly beyond normal experience. They are weird. But it does not then follow that every weird idea proposed is consistent with quantum mechanics. If quantum mechanics looks like magic, so do most great scientific advances when they first appear. As I have noted earlier, science begins with magic.

N. David Mermin succinctly summarized the conclusions of his own studies of the EPR paradox as follows: "If there is spooky action at a distance, then, like other spooks, it is absolutely useless except for its effect, benign or otherwise, on our state of mind" (Mermin 1985).

Further Reading

I will return later to other proposals for a quantum theoretical basis for psi phenomena not specifically based on the EPR paradox and Bell's theorem. For the reader who wants to learn more about the subject matter of this chapter, an excellent lay discussion of EPR, Bell's theorem, and the experiments designed to test them can be found in the *Scientific American* article by d'Espagnat (1979). I also recommend Mermin's article in *Physics Today* (1985). A more rigorous and very complete review is given in the *Physical Review* paper by Clauser and Shimony (1978). A full philosophical discussion of these issues can be found in the book by Redhead (1987), and the opinions on the subject of many prominent people have been collected by Davies and Brown (1986). An English translation of d'Espagnat's *Reality and the Physicist* has just become available (1989). Other skeptical critiques of paranormal applications of quantum theory have been given by Shore (1984) and Gardner (1983, 1985).

References

Aspect, Alain, Grangier, Philippe, and Gerard, Roger. 1982. "Experimental Realization of Einstein-Podolsky-Rosen-Bohm Gedankenexperiment: A New Violation of Bell's Inequalities." *Physical Review Letters* 49:91; "Experimental Tests of

Bell's Inequalities Using Time-Varying Analyzers." Ibid., p. 1804.

Bell, J. S. 1964. *Physics* 1:195.

Bohm, David. 1951. *Quantum Theory*. New York: Prentice-Hall.

———. 1952. "A Suggested Interpretation of Quantum Theory in Terms of 'Hidden Variables,' I and II." *Physical Review* 85: 166.

Bohm, D., Hiley, B., and Kaloyerou, P. N. 1987. "An Ontological Basis for Quantum Theory." *Physics Reports* 144(6):321.

Bohr, N. 1934. *Atomic Theory and the Description of Nature*. Cambridge: Cambridge University Press.

Clauser, John F., and Shimony, Abner. 1978. "Bell's Theorem: Experimental Tests and Implications." *Rep. Prog. Phys.* 41:1881.

Davies, P. C., and Brown, J. R., eds. 1986. *The Ghost in the Atom*. Cambridge: Cambridge University Press.

De Broglie, L. 1930. *An Introduction to the Study of Wave Mechanics*. New York: E. P. Dutton.

D'Espagnat, Bernard. November 1979. "The Quantum Theory and Reality." *Scientific American*. P. 128.

———. 1989. *Reality and the Physicist: Knowledge, Duration and the Quantum World*. Cambridge: Cambridge University Press.

Einstein, A., Podolsky, B., and Rosen, N. 1935. "Can the Quantum-Mechanical Description of Physical Reality Be Considered Complete?" *Physical Review* 47:777.

Einstein, Albert. April 26 and May 28, 1948. Letters to D. S. Mackey, as quoted in Fine, Arthur. "The Shaky Game: Einstein Realism and the Quantum Theory." In *Science and Its Conceptual Foundations*. David L. Hull, ed. Chicago: University of Chicago Press. P. 86

Gardner, Martin. 1983. "Parapsychology and Quantum Mechanics." In *Science and the Paranormal*. George O. Abell and Barry Singer, eds. New York: Scribners. P. 56; also in *A Skeptic's Handbook of Parapsychology*. 1985. Paul Kurtz, ed. Buffalo, N.Y.: Prometheus Books. P. 585.

Herbert, Nick. 1985. *Quantum Reality*. New York: Anchor Press, Doubleday.

Mermin, N. David. 1985. "Is the Moon There When Nobody Looks? Reality and the Quantum Theory." *Physics Today* 38:38.

Redhead, Michael. 1987. *Incompleteness, Nonlocality and Realism*. Oxford: Clarendon.

Sarfati, J. 1977. "The Case for Superluminal Information Transfer." *MIT Technological Review* 79, no. 5.

Shore, Steven N. 1984. "Quantum Theory and the Paranormal: The Misuse of Science." *Skeptical Inquirer* 9:24.

Targ, R., and Puthoff, H. 1977. *Mind-Reach*. New York: Delacorte Press.

Zukav, Gary. 1979. *The Dancing Wu Li Masters*. New York: Morrow.

11.

Thought Waves and the Energies of Consciousness

Crystals are amplifying minerals. You have a crystal in a radio—it amplifies the sound waves. You have a crystal in a television set—it amplifies the light waves. When you hold crystals, they amplify thought waves.

Shirley MacLaine

Animal Magnetism

In the 1770s, Franz Anton Mesmer stumbled upon the art of hypnotism. For a while the ability to gain apparent control over a subject's mind was referred to as "mesmerism." Mesmer himself thought he had discovered a new force, which he called "animal magnetism." He attributed it to an invisible fluid that connected the human body to the cosmos. The distribution of animal magnetism within the body determined a person's health and well-being.

No direct evidence was ever found for animal magnetism. Hypnotism is a real phenomenon, but one that does not require new kinds of forces or energies. Yet, the idea persists that some special type of energy exists in association with living organisms, and that this energy connects to mind and the cosmos.

It has long been a common belief that living organisms have some special spark of life, a vital force that distinguishes them from ordinary matter. We see this belief reappear in current questions over how to legally define the beginning and end of a human life. Let me make a prediction: these questions will never be answered, because they are unanswerable.

Defining life in a black-and-white fashion is impossible, because no single quality exists that distinguishes living from nonliving matter, animate from inanimate. Rather, the qualities of life, and the even more intricate qualities of mind, are sets of properties that develop as matter becomes increasingly complex, either through the natural processes of evolution or the artificial processes of the laboratory or factory. As we will see, the qualities we associate with life and mind exist in varying degrees in all systems of organized matter. The more highly organized the matter, the greater the number and variety of these qualities, and the higher the degree of development of life and mind.

The Human Energy Field

On February 13, 1973, the Second Western Hemisphere Conference on Kirlian Photography, Acupuncture, and the Human Aura was held in Town Hall, New York City. Over a thousand people listened to reports on experimental work that purported to demonstrate how human beings and other living things are surrounded by bioenergetic fields, or auras, and that these fields can be influenced by consciousness.

Most of the results presented at the conference involved the phenomenon of Kirlian photography. The reader has probably seen such pictures, which usually show fingertips or leaves surrounded by glowing streamers of light. Many examples and their fanciful interpretations can be found in the published proceedings of the New York conference, *The Energies of Consciousness: Explorations in Acupuncture, Auras, and Kirlian Photography,* edited by Stanley Krippner and Daniel Rubin (1975).

The Kirlian Aura

The Kirlian aura was discovered by an Armenian electrician, Semyon Davidovich Kirlian, in 1937. One day while repairing a high-voltage electrical device used in medical therapy, Kirlian experienced sparking between his hand and the device. Setting up an experiment, he obtained a photograph of his hand surrounded by a remarkable glow.

Kirlian and his wife Valentina continued to investigate the effect, hoping it might be of value in medical diagnosis (Singer 1981). They attracted little attention until 1970, when they were introduced to the West in the book *Psychic Discoveries Behind the Iron Curtain,* by Sheila Ostrander and Lynn Schroeder (1970).

The basic Kirlian experiment involved applying the outputs of a high frequency (75,000 to 200,000 Hertz) spark generator to two clamps, with the object to be photographed and the photographic film placed between the clamps. In addition to photographing the results, the Kirlians were also able to make live observations by using optical lenses to magnify the images. What they saw when a hand was placed between the clamps was a dynamic, changing pattern of multicolored sparks, twinkles, and flares— a veritable fireworks display.

Ostrander and Schroeder describe what happened when the Kirlians experimented with plants. A fresh leaf exhibited dynamic effects similar to the human hand, but a withered leaf displayed few flares, with sparks that hardly moved. "As they watched, the leaf seemed to be dying before their eyes and its death was reflected in the picture of the energy impulses. 'We appeared to be seeing the very life activity of the leaf itself,' the Kirlians said. 'Intense, dynamic energy in the healthy leaf, less in the withered leaf, nothing in the dead leaf' " (Ostrander 1970, p. 200).

The Kirlians found that inorganic objects, such as coins, showed only an even glow around the edges, while a living leaf or human hand had millions of sparkling lights "that glowed and glittered like jewels."

This remarkable "human energy field" was apparently sensitive to emotional and psychic states, at least according to Ostrander and Schroeder. "Illness, emotion, states of mind, thoughts, fatigue, all make their distinctive imprint on the pattern of energy which seems to circulate continuously through the human body" (Ostrander and Schroeder 1970, p. 207). The authors do not explain how this fantastic conclusion can be drawn from simple Kirlian photographs of fingers and leaves.

Kirlian Studies in the United States

Studies with Kirlian photography in the United States were pioneered by Thelma Moss and Kendall Johnson, who further developed the basic experimental technique. In the typical experiment, a high voltage is applied between the object and an aluminum plate (see Figure 11.1). A piece of unexposed photographic film is placed on the plate, separated from the plate by a dielectric isolator. The object whose aura is to be photographed is then pressed against the film. In order to prevent burns, the voltage is pulsed and the current carefully monitored. The experiments are really very simple, though care must be taken in the handling of any high voltages.

Moss and Johnson obtained many beautiful photographs, in color as well as black-and-white. However, they insisted that these images were more than simply a new art form. From their pictures of the human finger pad,

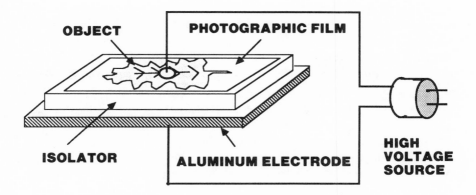

Figure 11.1. Typical apparatus for Kirlian photography. The object, in this case a leaf, is placed on a piece of photographic film which is electrically isolated from an aluminum electrode. Pulsed high voltage is then applied between an electrode placed in contact with the object and the aluminum electrode.

they claimed to "corroborate what the Soviet literature had reported, that different states of emotional or psychological arousal in the human being would reveal very different patternings in these photographs". (Moss and Johnson 1975).

This was, in the minds of Moss and Johnson, evidence for the existence of unseen fields associated with living matter: "All living things have not only a physical body, but also an 'energy body' consisting of 'bioplasma' " (Moss 1974).

Naturally, the media were fascinated by Kirlian photography—doubly so since stories could be accompanied by colorful eye-catching photographs. Had evidence at last been found for an energy field associated with living matter and human consciousness? Are our bodies and other living things surrounded by an aura that can be detected under appropriate experimental conditions? Can that aura provide us with practical benefits, such as early diagnosis of disease? Alas, once again, science has rained on the parade of paranormalists before the cameras.

Corona Discharge

Anybody with a modicum of physics training is bound to be underwhelmed by the Kirlian effect. The phenomenon of the Kirlian aura has been understood for a hundred years or more. It is simply the "corona discharge" that occurs when a high electric field is applied to a gas, in this case, air.

Above a certain threshold, the electric potential will cause a gas to break down to produce a violent, high current spark, such as lightning in the atmosphere. Below this threshold, a much less violent and lower current discharge of the gas takes place, as atoms are excited but not completely ionized. This is the corona discharge.

The first known image of a corona discharge was reported as far back as 1777, so it is really inaccurate to say that an Armenian electrician first observed it 160 years later. Audience-pleasing tricks with high voltages go back to the early days of applied electricity. In demonstrations that undoubtedly wowed his audience, the great inventor Nikola Tesla (1856–1943) would run high-voltage, high-frequency, low-amperage electric current through his own body, generating a halo about himself and causing sparks to shoot from his fingertips (Singer 1981).

No Redeeming Psychic Value

Physicists John O. Pehek, Harry J. Kyler, and David L. Faust have made a careful study of the Kirlian effect and published their results in the journal *Science* (1976). They and other experimenters have convincingly demonstrated that the effect is purely physical and without any redeeming psychic value.

Pehek and his coauthors showed that the most important variable in determining the type of image produced is simply the moisture at the contact between the subject and the film. The corona discharge is extremely sensitive to the amount of moisture, since this has a strong effect on the electrical conductivity of the air, and large variations occur as moisture is transferred from the subject to the film emulsion or simply evaporates. This alters the electric charge pattern on the film, resulting in different light streamers and sparks on the resulting photograph.

The Kirlians had claimed that only "living matter" exhibited a dynamic aura. Pehek, Kyler, and Faust disproved this, showing striking photographs of a piece of wood when dry, and then soaked with tap water. The dry wood has the uniform glow around the edges that is characteristic of "dead" objects, such as coins, while the wet wood shows "living" sparkles and streamers (see Pehek et al. 1976 or Singer 1981, p. 206).

Experiments exhibiting the effects of various physical parameters on Kirlian images have been done by University of Arizona physicists Arleen Watkins and William Bickel. They experimented with a number of variables: the type of photographic paper, pressure applied to the object, voltage discharge, exposure time, moisture, and developing time in photographs taken of the same finger.

While large variations are seen in their results, the amount of moisture is again easily shown to be the primary factor in producing the streamers that are claimed to be associated with living things (Watkins and Bickel 1989).

Watkins and Bickel also demonstrated the "phantom effect," whereby an aura is claimed to remain after an object, such as a leaf, is removed. They show that this is an artifact of the moisture, dust, and leaf juices left on the plate.

Cries of Agony

The widely acclaimed connection between the types of observed images and the emotional state of the subject, including the presumed "cries of agony" from a mutilated leaf, can be accounted for by moisture variations and is in no way correlated uniquely with living matter. Obviously, a leaf dries as it withers, and the pattern then shifts from the dynamic one associated with wet life to the static one associated with dry death.

As for possible sensitivity to human emotions in photographs of human fingers or hands, this phenomenon, if it can really be demonstrated to exist, can also be easily and naturally understood. This effect is most probably similar to that of the polygraph, or lie detector, in which moisture is one of the key parameters measured to decide whether the subject is telling the truth. We sweat involuntarily when we lie, but I would never bet my life on the accuracy of a polygraph. The lie detector has been shown to be a highly unreliable instrument and should never be used to determine guilt or innocence under any circumstance.

Actually, the Kirlian technique may be more accurate than a standard polygraph in its sensitivity to moisture, although this feature cannot be easily exploited because of the nonquantifiable nature of the observations. How would you apply it? Count sparks?

The various colors observed in Kirlian photographs can be shown to depend more on the type of backing used on the film than the emotional state of the subject (Singer 1981). And, for the final nail in the coffin of Kirlian photography, it is not observed in a vacuum. Surely something so cosmic as conscious energy fields cannot be stopped by a vacuum, like sound energy. Or turning it around, the fact that the Kirlian aura is so sensitive to simple physical parameters like moisture and air density can be taken as strong evidence that it too is natural rather than paranormal in nature. Surely true "spiritual energy" cannot depend on the number of N_2, O_2, and H_2O molecules present.

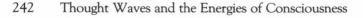
Our Natural Auras

Other means besides Kirlian photography can produce observable auras of light around objects, though not always so spectacularly as a Kirlian photograph. We all radiate light that is mostly in the infrared region of the spectrum and invisible to the eye, yet perfectly detectable with infrared sensitive devices and film. Photographic auras can occur from even more mundane causes, such as a dusty lens. The optics of the eye can also play tricks, sometimes giving the visual impression of auras around objects. Here's a simple experiment you can do right now: hold a finger two inches from your eye in front of a bright light. You will see your own aura.

Searching for the Psychic Mechanism

The remarkable thing about the Kirlian phenomenon is sociological, not physicial or psychical. How is it that so many people, including presumably highly educated parapsychologists, can be taken in by what, to any student of physics, is an almost trivial electrical effect that has been understood for a century? Electromagnetism holds no mysteries for the physicist, and no promise of being the cosmic fluid sought by the paraphysicist, psychic, or supernaturalist.

The classical electric and magnetic fields were described and unified with Maxwell's equations in 1873. As long as one stays out of the quantum regime, these equations still stand, confirmed and tested in countless ways. Electromagnetic quantum effects were described with great accuracy by quantum electrodynamics by the 1950s. The unification of the electromagnetic and weak nuclear forces was the physics triumph of the 1980s.

Mysteries still can be found at the unification level (though, as I have emphasized, no experimental anomalies currently exist), but these mysteries do not manifest themselves until we reach energies almost a billion times greater than the atomic processes that take place in normal living matter, including the human brain. Electroweak unification plays a role in the cosmos, but has effects on earth that can only be observed at our largest particle colliders or in the rare decays of particles not found naturally in great quantities on earth.

This has not stopped proponents of TM, for example, from asserting that the "cosmic consciousness" into which TM channels you is precisely the "grand unified field" of physics. Recall that the founder of TM, Maharishi Mahesh Yogi, has a degree in physics. So he should know what he's talking about.

A Quantum Mechanism?

We have already seen how some people have looked to quantum mechanics to provide a mechanism for communication channels beyond those of conventional physics. This has been shown to be based on a misunderstanding about what quantum mechanics really means. Nevertheless, the idea that quantum mechanics and modern quantum field theory somehow provide a framework for psychic phenomena remains the most commonly occurring theoretical idea in recent paranormal literature.

Electromagnetic waves are demonstrably not the vibrations of a cosmic aether and are considered a poor candidate for the psychic mechanism except among the scientifically illiterate. However, even scientifically sophisticated observers can be taken in by the deeper notion that the quantum mechanical wavefunction, still mysterious to some, carries the energies of consciousness.

Alleged psychic effects have no other known physical foundation. This is perhaps the reason why modern parapsychologists cling so desperately to quantum mechanics. It's all they have left. So let us return to quantum mechanics and the theories that purport to describe a connection between the mind and the wavefunction.

Statistical Determinism

We have seen that the conventional statistical interpretation of quantum mechanics is supported by experiments that rule out certain classes of alternate theories. We now can be almost certain that intrinsically real and local hidden variables cannot determine the outcome of events in the universe. Had the experimental tests of Bell's inequality by Aspect and others turned out differently, had they confirmed local reality, then the universe might have been restored to the kind of mechanical determinism that characterizes classical Newtonian physics. This did not happen.

Let me emphasize, however, that quantum mechanics did not totally destroy the notion that physical systems operate in a lawful manner. A form of statistical determinism still exists within conventional quantum mechanics. While individual particle behavior is unpredictable in the standard view, the statistical behavior of systems of particles remains calculable. If that were not so, then quantum mechanics would not be a scientific theory, since it could not be tested. For something to be science, it must be testable.

Probability

The statistical behavior of systems of particles is characterized by the average values of various measured quantities, and by the way those measurements are distributed. For example, suppose we measure the transverse positions x of a large number of electrons extracted in a pulse from an accelerator (Figure 11.2). We could plot the number of electrons found in various position intervals across the beam to obtain the bar graph shown. Dividing the number in each interval by the total, and connecting the points by a smooth curve, we would obtain the probability curve shown at the bottom.

Probability is simply the chance, the odds, that something occurs. Quantitatively, we define probability as the fraction of times that an event happens in a very large number of identical trials. For example, if I flip a coin one million times, it will come up heads 500,000 times, give or take a few hundred, so the probability for heads is 0.5. Chance fluctuations will make exactly 500,000 heads unlikely in any given million trial run, but as we increase the number of trials, the fraction of heads will get closer and closer to exactly 50 percent.

With our electron beam, we can directly identify the probability that an electron will be found at a particular position x with the frequency of occurrence of electrons in a position interval centered at x. For example, if the beam pulse contains 10,000 electrons, and 3,000 are found with a value of x between, say, 12.5 and 13.0 millimeters, then the probability of the electron falling in that range is $3,000/10,000 = 0.3$. If we view this as 10,000 separate experiments with one electron, rather than one experiment with 10,000 electrons, we would predict that the electron will, on the average, be found in the given range in 3,000 experiments and in 7,000 experiments will be found in any of the other ranges.

The most common probability curve applying to measurements of many types is the famous bell-shaped curve, called the Gaussian or normal curve, though many other shapes are possible. The Gaussian curve is characterized by two numbers: the average value of x, which is the same as the value at the peak of the curve in this case, and the standard deviation, or sigma, which is a measure of the width of the curve. Sometimes the standard deviation is referred to as the "error" in x, since it is a measure of how well the value of x is determined. If the standard deviation is small, most of the electrons will be found near the average position. If it is large, the electron positions are highly uncertain.

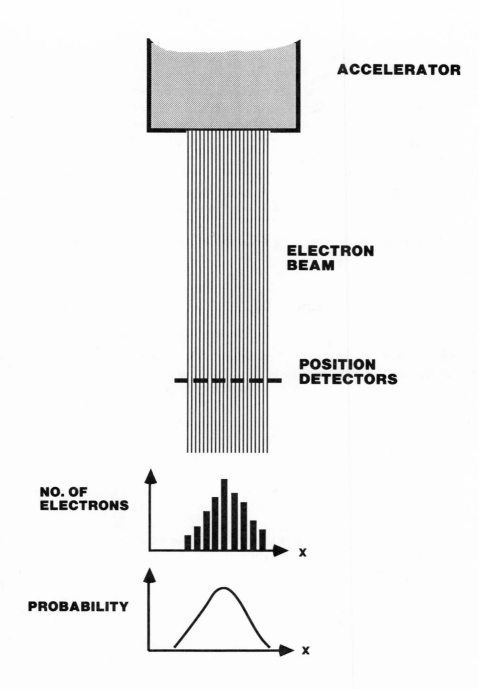

Figure 11.2. A beam of electrons from an accelerator passes through an array of position detectors. The bar graph indicates the number of electrons found at various positions. This can be converted into the probability curve shown.

Quantum Probability

In quantum mechanics, probability is proportional to the square of the wavefunction. In the typical quantum mechanics problem, we solve an equation, first introduced by Erwin Schrödinger in 1926, for the wavefunction. Squaring the result gives the probability for finding the particle within a unit volume, and so we can predict the average position of an ensemble of similarly prepared particles and the shape of its probability curve. The way in which the probability curve changes with time is calculable to any desired precision, at least in principle. So, in a sense, quantum mechanics is still "deterministic." The motion of individual particles is likely to follow this curve, but is not predictable with complete certainty. In the absence of hidden variables, the motion of individual particles is undetermined by preceding events.

Collapsing Wavefunctions

Now suppose that we place a screen with a small hole in it in our electron beam, so that only electrons directed through the hole can pass beyond the screen. This is equivalent to an act of measurement, since we now know the position of the electrons that appear beyond the screen to an accuracy equal to the diameter of the hole. When we measure the position distribution of electrons beyond the screen, we will get a much narrower curve, as indicated in Figure 11.3.

Since the square of the wavefunction still represents the probability curve, we say that the wavefunction "collapsed" in passing through the hole. By the very act of measuring the position of electrons in the beam, we have caused the wavefunction instantaneously to change from one whose square is given by the curve in Figure 11.2, to the one shown in Figure 11.3. The way in which the wavefunction collapses is not described by quantum mechanics at present; perhaps someday the theory will be extended to do this.

An Act of Consciousness?

This notion of instantaneous wavefunction collapse by virtue of the act of measurement has formed the primary basis of quantum theories of the paranormal. Because measurement is an act of human consciousness, it is supposed that, by measuring, the human mind is somehow able to control the behavior of particles.

This notion has been particularly expounded by Helmut Schmidt and

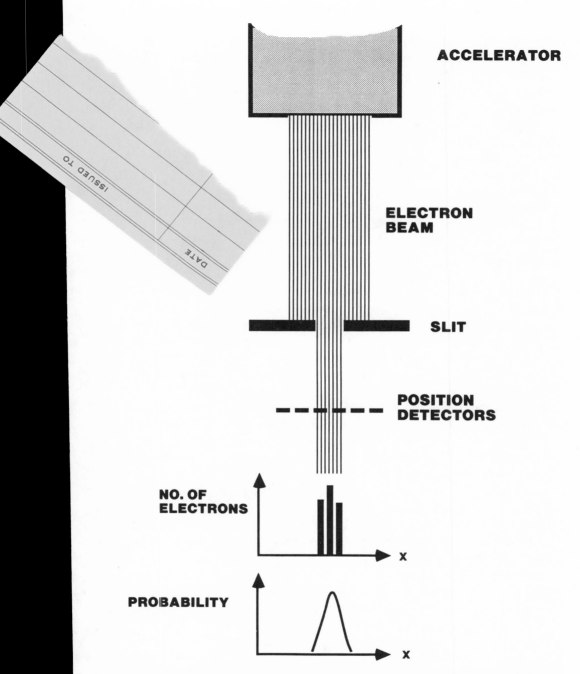

ACCELERATOR

ELECTRON BEAM

SLIT

POSITION DETECTORS

NO. OF ELECTRONS

x

PROBABILITY

x

ISSUED TO

DATE

Figure 11.3. The electron beam from the accelerator in Figure 11.2 passes through a narrow slit before the electrons have their positions measured. The resulting probability distribution is narrower. Thus the electron positions are determined with greater precision, as indicated by the narrower probability distribution. Since probability is proportional to the square of the wavefunction, the act of observing the position of the electrons causes their wavefunctions to collapse.

Robert Jahn, whose experiments with mental control of random number generators has been discussed earlier, and Evan Harris Walker. Considerable discussion can also be found in a number of popular books that evince a range of credulity on the part of the authors (see, for example, Targ and Putoff 1977, Zukav 1979).

Walker has even gone so far as to propose a mechanism by which consciousness can act on matter, providing an "explanation" for PK. We have already seen that PK remains to be experimentally demonstrated. But assuming that it exists, Walker tells us how it might happen. A detailed description of Walker's theory can be found in *The Iceland Papers* (Mattuck and Walker 1979).

Information

The key idea in Walker's quantum explanation of PK lies in the connection between probability and information. Information theory is a highly advanced discipline applied to signal and data processing in computers and communication electronics. It defines a mathematical quantity called information (I) as the negative of the base two logarithm of the probability (P) of a transition between two states, such as the binary bits in a computer memory ($I = -\log_2 P$, or $2^{-I} = P$).

For example, suppose we are attempting to communicate by means of a semaphore lamp or telegraph, using something like Morse code, but simpler. Let us translate that code into the binary bits used in computers and modern digital communications, with the short signal, or dot, represented by the value 0 and the long signal, or dash, represented by a 1.

Suppose we limit our communication to the twenty-six letters of the alphabet (spelling out numbers, for example). Assuming all the letters are equally likely (as will be the case for coded communication), then the probability for any one is $1/26$. From our definition, I equals minus log base 2 of $1/26$, which equals 4.7 ($2^{-4.7} = 1/26$).

In practice, $I = 4.7$ means that our code requires at least five bits (or five dot/dash combinations) to transmit the message of a single letter of the alphabet. Actually, with five bits we can send up to $2^5 = 32$ characters, or six more than in the alphabet. (Morse code is able to use fewer than five bits in most cases by not being strictly binary and utilizing pauses. It also transmits the numbers zero to nine, periods, and commas by using six or seven bits in a few cases.)

Getting back to the Walker theory of PK, since the brain processes information, and both information and the wavefunction have something to do with probability, Walker suggests that the brain has something to do with the wavefunction.

Moving Spheres

Mattuck and Walker consider the problem of moving a small sphere suspended in a liquid. They argue that the rate of information processed by the brain is about fifty bits per second per nerve fiber, multiplied by three million fibers, or about 150 million bits per second. Assuming that about 100 seconds is required for the sphere to be brought to rest by friction, they then say that the brain can provide fifteen billion bits of information to collapse the wavefunction of the sphere from some initial state of purely thermal motion to a final state where it is moving with a particular speed v_o.

Equating this information to the minus base 2 log of the probability expressed in terms of the initial and final state wavefunctions, Mattuck and Walker compute that the sphere would be set in motion at $v_o = 0.001$ centimeters per second. Unfortunately, this is far too slow to be observed.

Because their proposed theoretical mechanism does not produce measurable psychokinetic effects, Mattuck and Walker add further assumptions. Since 0.001 centimeters per second is too small an effect, they assert that the wavefunction is changed in 2,000 small steps to produce a measurable effect of $v_o = 2$ centimeters per second that they then claim is consistent with observations (Mattuck and Walker 1979).

The discerning reader can make his or her own evaluation of the sequence of steps from nerve fibers in the brain to collapsing wavefunctions to psychokinesis. The Mattuck and Walker paper is filled with impressive looking equations and calculations that give the appearance of placing psychokinesis on a firm scientific footing. I can see how lay persons could be misled into believing that science really supports PK.

Yet look at what they have done. They have found the value of one unknown number (wavefunction steps) that gives one measured number (the supposed speed of PK-induced motion). This is numerology, not science.

Although they do not say so explicitly, it appears that Mattuck and Walker are making the following assumptions: (1) individual particle behavior is controlled by the wavefunction, in violation of quantum mechanics; (2) the wavefunction is a real physical entity, perhaps the vibration of a cosmic aether; and (3) the brain interacts with the cosmic aether, producing waves which are transmitted through space at speeds greater than the speed of light, providing for instantaneous wavefunction collapse.

None of these assumptions is supported by the evidence. In pilot wave theories such as those of Louis de Broglie and the more modern versions of David Bohm, some physical reality is attributed to the wavefunction. But these are not required by any known empirical data. By the law of parsimony, conventional quantum mechanics still reigns, with the wavefunction interpreted as a purely mathematical entity whose square represents the probability that certain events occur.

Nothing to do with Consciousness

If the act of measurement causes the collapse of the wavefunction, this still need have nothing to do with human consciousness. The collapse of the wavefunction happens with impersonal automatic detecting equipment just as assuredly as when a person is watching.

The example that is usually raised at this point has become trite, but I must respond to it anyway because it is often thrown at me when I lecture. Does a tree that falls in the forest, with no one present to hear it, still make a sound? How can there be a sound without an ear to detect it and without a human consciousness to understand it? Well, the sound could just as well be picked up on a tape recorder left nearby. It would be nonsense to say that the tree fell only when the tape was later played back to a human listener.

Suppose the tape recorder is removed, and the forest later burns down before the tape is played for any listener. Would the burning of the forest wipe out the tape record of the falling tree? Paranormal thinking would say that it would, since the tree does not fall until a human hears the tape, and the forest has burned down in the intervening time. But do you really believe that this happens?

The paranormal prediction can easily be tested by actually performing the experiment. Paranormalists, get out your matches! But please don't burn down a whole forest. Just a small laboratory model will suffice.

Abrupt Probability Switch

The collapse of the wavefunction expresses nothing more than the fact, well known from probability theory, that the mathematical probability of an event abruptly switches to unity once that event occurs.

For example, suppose you have bought a ticket for a state lottery for which twenty million tickets were sold. Prior to the selection of the winning number, the probability for winning was one in twenty million.

The moment that the winning number is announced, however, the probability for the winner instantaneously switches to unity, while the probability for the remaining ticket holders instantaneously switches to zero. So, when the winner is announced, every player's wavefunction instantaneously collapses.

If some of the twenty million ticket holders are located at geographically distant points, it will take several microseconds or more for the news to reach them via telephone, radio, or TV. Does a violation of Einstein's theory of relativity occur, with information traveling faster than the speed of light? Of course not.

Suppose you are 300 kilometers away from where the drawing takes place, but you are watching it on TV. While your purely mathematical probability of winning—if you lose—will go to zero at the instant the number is called, you actually will not know about it until at least a thousandth of a second later, when the TV signal reaches you (even longer if it is beamed up to a satellite and back down).

In other words, the information flow from the TV camera to your set still moved no faster than the speed of light, consistent with relativity. Your mathematical probability may previously have gone to zero, but the useful knowledge of that fact—the actual physical signal transmission that contained the information—occurred in a manner totally consistent with the known principles of physics.

The confusion here is similar to that with the EPR paradox. As we have seen, the wavefunction in the EPR experiment only affects the statistical distribution of observed photon polarizations. This can carry no new information and so cannot be used for signal transmission. Even when the wavefunction collapses faster than the speed of light, signals containing knowledge of that fact can move no faster than the speed of light.

No Cosmic Fluid

So no evidence exists that human consciousness is connected in any way to an all-pervading cosmic fluid, through electromagnetic aural waves or quantum mechanical particle waves. To the best of our knowledge, the universe is composed of discrete chunks of matter that interact locally. The seeming nonlocal action of quantum mechanics occurs only for abstract quantities, such as the wavefunction, which are figments of the physicist's imagination. The wavefunction does not correspond to any directly measurable entity, so its nonlocality cannot be interpreted as a force that acts at a distance in violation of relativity.

After more than a half century, the conventional interpretations of relativity and quantum mechanics remain completely consistent with every observation we have made about the universe. None of these observations or the theoretical constructs that so beautifully describe the observations provides any signal of a world beyond matter.

As for quantum mechanics providing a basis for our anthropocentric delusions of grandeur, Martin Gardner has remarked: "If you are compelled to think, for emotional reasons or because some guru said so, that you are essential to the universe, that the Moon would not be there without our minds to see it (the mind of a mouse? Einstein liked to ask), you are welcome to such self-centered insanity. Don't imagine it follows from quantum mechanics" (Gardner 1989).

References

Gardner, Martin. 1989. "Guest Comment: Is Realism a Dirty Word?" *American Journal of Physics* 57 (3):203.

Heisenberg, Werner. 1959. "Planck's Quantum Theory and the Philosophical Problems of Atomic Physics." *Univeritas* 3, no. 2. Reprinted in *Science: Men, Methods, Goals.* 1968. Brody, Boruch A., and Capaldi, Nicholas, eds. New York: W. A. Benjamin, 1968.

Krippner, Stanley, and Rubin, Daniel, eds. 1975. *The Energies of Consciousness: Explorations in Acupuncure, Auras, and Kirlian Photography.* New York: Gordon and Breach.

Mattuck, Richard D., and Walker, Evan Harris. 1979. "The Action of Consciousness on Matter: A Quantum Mechanical Theory of Psychokinesis." In *The Iceland Papers.* Puharich, Andrija, ed. Amherst, Wisc.: Essentia Research Associates. P. 111.

Moss, Thelma. 1974. *The Probability of the Impossible.* Los Angeles: Tarcher.

Moss, Thelma, and Johnson, Kendall. 1975. In Krippner and Rubin. 1975. P. 41.

Ostrander, S., and Schroeder, L. 1970. *Psychic Discoveries Behind the Iron Curtain.* Englewood Cliffs, N.J.: Prentice-Hall.

Pehek, John O., Kyler, Harry J., and Faust, David L. 1976. "Image Modulations in Corona Discharge Photography." *Science* 194:263-270.

Singer, Barry. 1981. "Kirlian Photography." In *Science and the Paranormal.* Abell, George O., and Singer, Barry, eds. New York: Scribners. P. 196.

Targ, R., and Puthoff, H. 1977. *Mind-Reach.* New York: Delacorte Press.

Watkins, Arleen J., and Bickel, William S. 1986. "The Kirlian Technique: Contolling the Wild Cards." *Skeptical Inquirer* 10-3:244.

———. 1989. "A Study of the Kirlian Effect." *Skeptical Inquirer* 13-3:172.

Zukav, Gary. 1979. *The Dancing Wu Li Masters.* New York: Morrow.

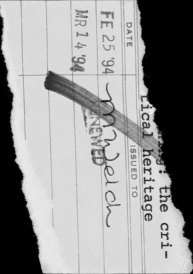

12.

The Broken Whole

If we are ever to discover the laws of nature, we must do so by obtaining the most accurate acquaintance with the facts of nature, and not by dressing up in philosophical language the loose opinions of men who had no knowledge of the facts which throw most light on these laws. And as for those who introduce ethereal or other media to account for these actions, without any direct evidence of the existence of such media, or any clear understanding of how the media do their work, and who fill all space three and four times over with aethers of different sorts, why the less these men talk about their philosophical scruples about admitting action at a distance the better.

James Clerk Maxwell

Lodge's Last Salvo

We have already met physicist Oliver Lodge, one of the handful of prominent scientists who involved themselves in the study of psychic phenomena in the latter part of the last century. His active scientific career spanned the period just prior to the revolution of twentieth-century physics. His work on wireless telegraphy helped lay the foundation for that revolution. Yet, by the time it had become clear to most physicists that a change in our knowledge of the basic nature of the universe had taken place, Lodge was too old to change the perspectives of a lifetime.

Instead Oliver Lodge joined an aging group of scientists who, to the end, refused to accept the full implications of the discoveries by the team of young whippersnappers led by Einstein, Bohr, and Heisenberg. Only with the help of the Grim Reaper did the new physics finally gain the consensus it enjoys today.

In his presidential address to the British Association for the Advancement of Science in 1913, entitled "Continuity," Lodge fired the last salvo for the departing troops of the old physics. In this speech, he deplored "the modern tendency . . . to emphasize the discontinuous or atomic character of everything" (Lodge 1914). Seven years later, in his book *Beyond Physics,* Lodge reaffirmed his lifelong belief in the "one all-embracing reality on the physical side, the Ether of Space . . . in terms [of which] everything else in the material universe will have to be explained" (Lodge 1920, pp. 45-46). Despite Einstein and the theory of relativity, Lodge still held fast to the concept of a continuous universal aether as the cosmic glue without which "there hardly can be a material universe at all" (Lodge 1914).

Apples, Planets, and Atoms

Newton's laws of motion applied to visible bodies such as apples and planets, but they were also descriptive of the atomic view of matter. Atoms had been postulated by the ancient Greeks, and a belief in the discrete nature of matter was widely held at Newton's time. Even light was thought by Newton to be atomic, or "corpuscular," in nature.

Many properties of matter were explained by the atomic model. Yet it took until early in this century to produce sufficient evidence to establish with certainty the size of atoms or the number of them in a gram of matter. As long as these deficiencies of the atomic theory remained, other possibilities could be legitimately entertained. One such possibility was that matter was a continuous fluid.

Continuity

To the naked eye, both matter and light appear smooth and continuous. It does not seem from everyday experience that the world is composed of tiny particles, with mostly empty space in between. Only with modern instrumentation can you see, that is, detect, individual atoms of matter or the discrete quanta of light.

In the time between Newton and Einstein, evidence accumulated for additional physical realities that also appeared to be continuous rather than discrete: gravity, electricity, and magnetism. Furthermore, these realities were invisible, detectable only by their effects, and possessing properties that, in a flight of imagination, one might extrapolate to a connection between the human body, the mind, and the cosmos.

Oliver Lodge's physics career occupied the post-Newtonian period when the notion of universal continuity seemed to be supported by the preponderance of evidence. Powerful new mathematical techniques had been developed and Newton's discrete particle mechanics was elaborated to encompass a description of continuous fluids.

This new "continuum mechanics" was applied with great success to the rapidly growing technology of the eighteenth and nineteenth centuries, from hydraulics to thermodynamics and optics. The engines of the industrial revolution were built, in large measure, on the theoretical framework of continuum mechanics.

Since matter appears continuous on the macroscopic scale of everyday experience, most common physical phenomena can be accurately described in continuum terms. The properties of fluids can be modeled by introducing the idea of continuous fields, such as pressure and density. These fields are not confined to one point in space, as particles are, but extend throughout space. The movement of the fluid is then calculated from equations of motion for the continuous fields.

A sound wave, for example, can be described in terms of oscillating pressure and density fields in which disturbances at one place can move through the medium to another place. Continuum mechanics predicts how an extended material system behaves under a specified impetus, such as the pressure impulse that generates a sound wave.

Like Newtonian particle mechanics, continuum mechanics was a completely deterministic theory. Every event resulted from the causal action of a preceding one. One of the many paranormalist misrepresentations of modern physics is that Newtonian determinism was destroyed by the development of theories of continuous fields. They would have us think that the theory of continua is something new, part of the grand new holistic paradigm of twentieth-century physics. This is simply false. As we see, it is centuries old.

The Holistic Paradigm

The success of continuum mechanics, in particular the theoretical concept of fields, as well as the absence at that time of direct evidence for atoms, provided nineteenth-century physicists like Oliver Lodge with a perfectly rational basis for the rejection of the atomistic, reductionist view of nature. If Einstein and his corevolutionaries had not come along, today's paranormalists might still argue that the atomic model should be replaced by a new universal holistic paradigm—a universal cosmic field—in which everything is a part of everything else. The holistic paradigm is perfectly consistent with physics—nineteenth-century physics. But, unfortunately for the holistic paradigm, Einstein did come along.

The Aether

Although Isaac Newton had promoted the corpuscular view of matter and light, he had also discovered principles that did not fit within a conceptual framework of discrete objects interacting locally. One of the puzzles that deeply disturbed his critics, and indeed Newton himself, was how gravity appeared to be a continuous, invisible force that acted instantaneously over cosmic distances.

Newton taught us how to calculate the gravitational force and use it to predict the motion of localized bodies, such as planets, with remarkable accuracy. But his equations did not tell us what gravity really was, or how this mysterious force was transmitted from one place to another.

At first Newton did not care, saying: "I frame no hypotheses." He was satisfied, as are most scientists when their equations work. His laws of gravity and motion were stupendously successful. They predicted the motions of planets and projectiles. His friend Edmund Halley used Newton's laws to predict the reappearance of a comet, now identified with his name, seventy-six years later. The appearance of Halley's comet on schedule in 1759, after Newton's and Halley's deaths, was a sensation that solidified public recognition of the great predictive power of science.

From a purely utilitarian perspective, nothing beyond computational power was needed. In the spirit of Occam's razor, why frame any unnecessary hypotheses not required by the data and not adding anything to the capabilities and applicabilities of the theory?

But like the Nobel laureate of today who appears on a talk show and finds himself asked questions that neither he nor anyone can answer, Newton was pressed for explanations. So, despite his better instincts, he eventually speculated on *why* his theories worked.

In the "Queries" that he added to his *Optiks* in 1704 and afterwards, Newton tried to explain gravity in terms of an unseen component of the universe called the "aether." Aristotle had introduced the aether as the fifth element, the stuff of the heavens. Air, earth, fire, and water were the grubby stuff of the world below the heavens.

Natural Motion

However, it was not the authority of Aristotle that led Newton to the aether, but rather an important break he made with that authority. Aristotle had said that the most perfect motion was circular. He considered circular motion natural to the heavens where only perfection was possible.

Newton declared that truly natural motion was straight-line motion, whether in heaven or on earth. This was indeed a revolutionary idea. Even

Newton's brilliant contemporaries, Descartes and Huygens, clung to the Aristotelian view of the primacy of circular motion.

Equally important to humanity's evolving concept of the universe, Newton announced that the laws determining the movements of bodies on earth likewise applied to the bodies of the heavens. This had to be if the earth was just another planet revolving about the sun and not the center of the universe.

A generation earlier, Galileo had looked through his telescope and confirmed the Copernican idea that the earth revolved about the sun, presumably in a circle. If, as Newton now said, straight-line motion is the only truly natural motion, then an invisible force of gravity is needed to cause the earth to deviate from its natural straight-line path.

You can see the problem with Newton's new theory: if all heavenly bodies move in circles, and circular motion is natural, then no invisible forces acting over cosmic distances are needed. Aristotle's view, seconded by Descartes and Huygens, made sense. Furthermore, it was economical, as required by Occam's razor. In abandoning the idea that circular motion is natural, Newton was forced to the uneconomical expedient of introducing the force of gravity as a mysterious, invisible entity.

But, there was no returning to natural circular motion, at least until Einstein's general theory of relativity. The planets do not move in exact circles, anyway. Kepler had shown that their paths are actually elliptical, an observational fact that Newton, in triumph, could derive from his theory. His theory was economical after all.

So Newton introduced the force of gravity to invisibly act to make planets fall around the sun and bodies fall to the earth. His theory worked superbly. But when he felt compelled to seek an explanation for gravity, Newton resurrected Aristotle's aether, speculating that it transmitted gravitational forces from point to point. He also suggested that this medium might account for electricity, magnetism, light, radiant heat, and the motion of living things that he, like his contemporaries, believed required some special principle.

Newton even proposed that the vibrations of the aether might be excited by the brain. These reluctant speculations about an invisible cosmic fluid by one of the greatest thinkers in history form the archaic conceptual foundation of most modern psychic theories.

Electromagnetic Fields

By the mid-nineteenth century, considerable data had been accumulated on the twin phenomena of electricity and magnetism. The great pioneer of electromagnetism, Michael Faraday, after some initial reluctance to speculate that mirrored Newton's, came to the view that space contained invisible fields under tension. He imagined lines of force, like invisible wires, emanating from bodies and merging with the field lines of other bodies to generate the observed forces between the bodies.

Faraday's lines of force provide a visual image of gravitational, electric, and magnetic fields. In a common classroom demonstration, iron filings sprinkled on a sheet of white paper placed over a magnet trace out the magnetic field lines. The density of lines of force at a given point in space—indicated by the density of accumulated iron filings—provides a measure of the strength of the field at that point.

The concept of invisible fields pervading space is often appropriated by those who, with little real knowledge of physics, attempt to make psychic phenomena seem scientific. I have already laid to rest the Kirlian aura. Still, you merely need to look at the covers of the books in the "Psychic Science" section of your bookstore to see artistic renderings of coronas, auras, and field lines emanating from human heads and reaching out to the cosmos. The idea of the field provides a vague mechanism for the cosmic consciousness.

In 1858, William Thomson proposed a theory of electromagnetism based on the mechanics of continuous fluids. This inspired another Scotsman, James Clerk Maxwell, to apply his considerable mathematical skills to the same problem. In 1873, Maxwell published his *Treatise on Electricity and Magnetism,* containing the equations, later reduced to only four, that are solved for the electric and magnetic fields. These equations encompassed laws previously discovered by Coulomb, Ampere, and Faraday, along with an added term. This term allowed for electric and magnetic fields to regenerate each other in the absence of any electric charges or currents, producing an oscillating electromagnetic wave.

Solving for these waves, Maxwell predicted that they should travel at the same speed as light. The implication was that light is a form of electromagnetic wave. Other forms of electromagnetic waves were also inferred, and in 1888 this prediction was confirmed when Heinrich Hertz produced and detected radio waves in the laboratory.

Although Maxwell himself had cautioned that his equations were simply mathematical abstractions, others, including Thomson and Hertz, believed they had to represent some sort of real physical entity. Light was an electromagnetic wave. Sound waves are vibrations of air or other materials, so it made sense that light must also be some kind of vibration of a medium.

But since light, unlike sound, can reach us from the stars, the vibrating medium that produces light must be invisible, frictionless, and continuously filling all of space. It seemed reasonable, in fact rather obvious, that electromagnetic waves were the vibrations of Aristotle's and Newton's celestial element: the aether.

Einstein Kicks Down the Pillars of Continuity

The idea that electromagnetic waves are vibrations of the aether lasted only a generation. In three immortal papers published in 1905, Einstein destroyed the aether and kicked down the remaining pillars of continuity.

First, in a publication based on his Ph.D. thesis for the University of Zurich, Einstein provided the equations that would enable measurements of Brownian motion to set the scale for atoms. The botanist Robert Brown had observed this chaotic motion of tiny particles in a fluid in 1827. Brownian motion is caused by atomic and molecular bombardment. With the scale for atoms finally determined, the last effective roadblock to the acceptance of the atomic theory of matter was removed.

Second, with his photon theory of light, based on the quantum of Max Planck, Einstein showed light to be discrete rather than continuous. And third, Einstein's special theory of relativity eliminated the need for an aether.

Not Aethereal but Ephemeral

We now have established that the atoms and molecules of matter and their nuclear and subnuclear constitutents do not interact with one another by invisible fields. Rather they exchange photons and other particles we call gauge bosons.

The fields and field lines of nineteenth-century physics are no longer part of the structure of fundamental physics. Although field lines are still used as a convenient visual tool, they exist only in equations and sketches on a chalkboard. They are mathematical abstractions, just as Maxwell suggested. Not aethereal but ephemeral, they disappear into oblivion when the board is erased.

Today, continuum mechanics survives as a useful approximation that works fine for many problems on the macroscopic scale, like the study of fluids, where the number of atoms is so large that the fundamental discreteness of matter and energy can be ignored. Airplane wings can still be aerodynamically designed without quantum mechanics.

However, the universe is fundamentally divided into parts. We have found that even space and time are ultimately discrete, though at such a small scale that they can be treated as continuous variables in most physics equations, including those currently used to describe events at the subnuclear level.

Oliver Lodge and his fellow champions of continuity are long gone. Yet the universal fluid idea lives on as the model used by proponents of the paranormal, at least implicitly, to justify their fantasy that the universe is one unbroken whole and that mind is somehow able to communicate with all of existence in an instantaneous, holistic way. But the arguments for this belief are far weaker today than they were in the time of Oliver Lodge.

The Heyday of Particle Physics

In the early days of my physics career, Berkeley, California, was the world center of particle physics. I had many occasions to visit and work there in the 1960s and early 1970s. Indeed, my Ph.D. thesis at the University of California at Los Angeles was based on an experiment done at Berkeley. After graduating, I continued to make many visits and spend pleasant summers working at the Lawrence Radiation Laboratory (now Lawrence Berkeley Laboratory), in the hills above the campus of the University of California.

The "Bevatron" accelerator at the Lawrence Lab was capable of producing protons with an unprecedented energy of six-billion electron-volts. It had been designed to produce the antiproton, the antimatter partner of the proton. By 1955, few physicists doubted the antiproton's existence, and its predicted properties were well established, though no antiprotons had yet been identified in the laboratory. So there was no surprise when antiprotons were found that year in the initial Bevatron experiment, the first experiment with sufficient energy to produce them.

With the addition in the 1960s of powerful new detectors like the bubble chamber, Berkeley began churning out major discoveries almost every week. A continual stream of new short-lived particles, called "hadrons," filled the journals. These hadrons did not fit into the then-existing picture of fundamental particles.

Theorists tried to make some kind of sense of the many new particles coming from Berkeley, and from other high energy particle accelerator centers that quickly sprang up in Europe and elsewhere in the U.S. Most researchers approached the problem of classifying these particles according to the traditional method of the physics and chemistry of the past. They assumed that such a great number of particles could not all be elementary, that a new layer of matter must be hidden underneath them.

Clues were sought as to what the next layer of fundamental constituents might be. Particle physicists were kept happily busy combing great reams of data for those clues. In this quest, we had all the support needed from our universities and the federal government. After all, fundamental physics had produced the nuclear bomb. National survival dictated that basic research into the structure and forces of nature receive highest priority.

During this time the unprecedented power of the new digital computers was unleashed. When a vastly superior new-generation computer was built, the first one out of the factory went to Berkeley, before the manufacturer's software was available. Berkeley had the programmer power to write a complete software system, and could do so faster than was possible anywhere else.

It was the heyday of particle physics, a simpler time that we can only look back to now with nostalgia.

Particles Galore

For quite a while we could not make much sense of this great array of new particles. Back in the 1930s, the discovery of the neutron had greatly simplified our view of the atom; the whole Periodic Table of the chemical elements was reduced to just three particles: electron, proton, and neutron. But then other, new particles, such as the muon, pion, and kaon, were seen in cosmic rays. Now, with the addition of numerous types of hadrons from accelerators, we were overwhelmed. Soon we had hundreds of particles, and no theory to explain them.

S-Matrix Theory

A novel approach occurred to a few theorists. Led by Geoffrey Chew of the University of California at Berkeley, they suggested that perhaps there were no elementary constituents at all. Or, alternatively, maybe we should regard all particles as equally elementary. They dubbed this theory "nuclear democracy," but it became more commonly referred to as the "bootstrap theory" or "S-matrix theory." Perhaps we at last had seen the end of atomism. A new holistic paradigm might prove the answer.

The bootstrappers were motivated by certain remarkable properties of the mathematics of complex variables, which are variables that contain both real number and imaginary number components (an imaginary number contains a factor of the square root of minus one). The S-matrix was a table of complex numbers that described the scattering of particles.

The hope was that from just a few very general and reasonable assumptions about the S-matrix, the properties (masses, spins, charges) of all particles would appear as the only self-consistent solution. That is, everything known about the particles would be calculated without any experimental data used as input to the calculation. The theory would stand alone, suspended in space by its own power and beauty with no grubby experimental facts needed to prop it up. Mother Nature, like Mammy Yokum of the old comic strip "Li'l Abner," pulls herself up by her own bootstraps. It was a wonderful thought.

The Quark Theory

While the S-matrix theorists struggled mightily but unsuccessfully to find the magical equations that explained everything, other theorists were following the more conventional approach of peeling away the next layer of matter in search of an underlying substructure. In particular, Murray Gell-Mann and George Zweig showed that they could explain many of the properties observed for the hadrons, as well as the proton and neutron, in terms of basic constituents that Gell-Mann dubbed "quarks."

By the early 1970s, experimentalists had accumulated evidence supporting the quark model. Thus, the way was paved for the ultimate development of the highly successful Standard Model that now explains all the observed properties of matter at the fundamental level.

Today, with the Standard Model firmly established, the ancient picture of the universe as a composite structure built up from a small number of basic elements has been retained, at least for the layer of matter now accessible to our best instruments. S-matrix theory has been banished for its failure to produce results, as well as because of the success of the quark model. If the bootstrap idea is ever to work, it will have to be within the world of quarks and leptons, or their constituents in an even deeper layer of matter—not protons, neutrons, and the hadrons that came out of the Bevatron and its sister accelerators.

Talk

Of course there was more to Berkeley in the late 1960s and early 1970s than hadrons and S-matrices. If the University of California at Berkeley was the center of the new particle physics, Telegraph Avenue, just off campus, was the center of the student antiwar movement and a focal point of many of the other movements we associate with those turbulent times. Out of that remarkable period came hippies, psychedelics, flower power, and the "new consciousness" that has now evolved to the New Age.

The Beatles had gone to India and discovered Eastern mysticism, bringing Maharishi Mahesh Yogi and his TM to the West. It became quite fashionable to wear Indian dress, burn incense, listen to sitar music, meditate, and of course, talk. . . .

The coffeehouses of Berkeley and other campus towns across the country were filled with talk: about the paranoia of the military mind and the dangers of nuclear war; about the greed of industry and environmental pollution; about inequity and injustice in the world, overpopulation, racism, and sexism; and about how science got us all into the mess we were in.

The University of California's two Lawrence Radiation Laboratories were renamed the Lawrence Berkeley Laboratory and Lawrence Livermore Laboratory, so people might not associate them with dangerous nuclear radiation (though Livermore continues to this day to be the primary U.S. nuclear weapons research center).

After years of privilege, it became a tough time to be a scientist. I found myself working "relevance" into my classes, to try to convince students that science was still a worthwhile endeavor, and to keep the enrollment from entirely slipping away, as students preferentially chose courses in ecology and ethnic studies. I presented material on physics applications in medicine and environment. To show that physics wasn't all military applications, in my example problems I dropped CARE packages—instead of bombs—from airplanes. More daringly, and not too successfully, I tried to indicate the intellectual and artistic adventure of the new physics, relativity, and quantum mechanics.

The Tao of Physics

One who tried a little harder to reconcile physics with the new consciousness was a young theoretical physicist from the University of Vienna, Fritjof Capra. Capra worked on S-matrix theory with Geoffrey Chew during the raucous days in Berkeley and saw wider applications of the holistic approach to understanding nature. Also exploring fashionable Eastern philosophy, he discovered what he thought were remarkable parallels between much of its teachings and modern physics, especially quantum mechanics. Capra also found much in common with his own speciality of S-matrix theory. These revelations he collected in a book called *The Tao of Physics* (1975).

The common thread Capra claimed to see tying Eastern philosophy to modern physics was "the unity of all things." According to Capra, the most important characteristic of the Eastern world view is that "all things are seen as interdependent and inseparable parts of the cosmic whole." This whole is called *Brahman* in Hinduism, *Tao* in Taoism, *Dharmakaya* or *Tathata* in Buddhism. The last is translated as "suchness," as in the quotation: "What is meant by the soul as suchness, is the oneness of the totality of all things, the great all-including whole" (Ashvaghosha 1900).

Capra found that the Chinese concept of *ch'i* resembled the physics concept of the field: "The Great Void cannot but consist of ch'i; this ch'i cannot but condense to form all things; and these things cannot but become dispersed so as to form (once more) the Great Void" (Chang Tsai, as quoted in Fung 1958).

In the vast literature of Eastern thought, surely something must have been said about almost everything and everything said about something. Capra finds a parallel to the quantum mechanical principle that the act of measurement cannot be separated from the phenomenon being measured. He makes the same connection we have seen others make, that quantum mechanics somehow implies a nonsensory interaction with the human mind, and that perhaps everything is just a creation of mind.

Capra quotes from the Yogacara school of Buddhism: "Out of mind spring innumerable things. . . . These things people accept as an external world. . . . What appears to be external does not exist in reality; it is indeed mind that is seen as multiplicity; the body, property, and above-all these, I say, are nothing but mind" (Suzuki 1952).

This is ancient Indian idealism, which we have already noted must have at one time trickled through the Middle East to Greece, where it probably influenced much of Western thought, including Christianity. The idea that we make our own reality is a major component of New Age thinking. Like everything else in the New Age, it is nothing new.

The Dancing Wu Li Masters

Similar connections between modern physics and Eastern mysticism were popularized by Gary Zukav in *The Dancing Wu Li Masters* (Zukav 1979). The book jacket calls it a "bible . . . for those people who have heard of the mind-expanding, psychedelic aspects of advanced physics." Actually, it's not that bad. The author, who is not a scientist, does a better job of explaining the new physics and conveying its wonder, beauty, and intellectual challenge than most scientists have been able to do. And there's little, if anything, "psychedelic" about the picture Zukav paints.

Wu Li, according to Zukav, is the English transcription for five different Chinese expressions, all of which he finds descriptive of the new physics:

Wu Li = Patterns of Organic Energy
Wu Li = My Way
Wu Li = Nonsense
Wu Li = I Clutch My Ideas
Wu Li = Enlightenment

I can see a little of physics in each of these myself. Zukav's Wu Li Masters are the great physicists, like Einstein, who dance with the rest of us, their students. They lead, but do not teach. The student learns from watching and following the master.

Zukav sees many of the same parallels between the new physics and Eastern mysticism that Capra sees: emptiness is form, reality is everything you can think of, and all existence is an unbroken wholeness. But at least Zukav is willing to consider alternative interpretations. In a vast improvement over Capra's random meanderings through Eastern literature to find a catchy quotation here and there that vaguely sounds like the new physics, Zukav makes a serious and quite successful stab at explaining how the new physics differs from the common person's view of science.

In particular, Zukav gives a very sound exposition of the EPR paradox, Bell's theorem, and the experimental situation prior to the published work of Aspect and his collaborators. All these earlier experiments had indicated what the Aspect group has now confirmed: that Bell's inequality is violated by exactly the amount predicted by conventional quantum mechanics.

Zukav carefully lays out the various interpretations of this result, wisely choosing not to follow the lead of the writers who immediately jump to the conclusion that superluminal connections must occur. Unfortunately, he is still a bit too credulous of the superluminal interpretation of psychic phenomena. Like so many popular writers, Zukav was bamboozled by the numerous oversold and undersupported claims that such effects have been observed, and that they act "instantaneously, if not faster" (Zukav 1979, p. 316).

Other Interpretations

But at least Zukav does admit that other interpretations of Bell's theorem are possible that do not require the violation of Einsteinian locality and superluminal connections. As I have previously noted, the Copenhagen Interpretation of quantum mechanics provides an acceptable interpretation by carefully defining what we mean by the "reality" of physical variables in quantum mechanics.

The Copenhagen Interpretation is the most economical solution, since it does not require additional hypotheses to those already present within the existing framework of conventional relativity and quantum mechanics. Why introduce superluminal connections that violate well-established principles, when you can use an interpretation already validated by all known data? Zukav discusses other possibilities that strike me as rather far-fetched, but at least he gives them all a fair shake.

In all, *The Dancing Wu Li Masters* is much better than the book jacket would make you think, where the book is praised because it is "without equations and without the 'scientific mentality.' " It's too bad that publishers feel they have to pander to an antiscientific element, even when their authors do not. Zukav avoids equations, but he does not disparage the "scientific mentality." In fact, he clearly admires science and scientists, and uses respectable, rational arguments throughout.

The S-Matrix and the Cosmic Way

It was to the S-matrix that Fritjof Capra finally turned in *The Tao of Physics* for the strongest connection with Eastern ideas. He relates the S-matrix to the cosmic way—the Tao. It is not laid down by any lawgiver, but is inherent in nature. The universe is the way it is, because it is the only way it can be. Quoting from the *Tao Te Ching* of Lao Tzu (Ch'u Ta-Kao 1973):

> Man follows the laws on earth;
> Earth follows the laws of heaven;
> Heaven follows the laws of *Tao;*
> *Tao* follows the laws of its intrinsic nature.

But, as I have said, S-matrix theory failed to provide a useful description of elementary particles. The effort to find a self-consistent set of equations that predict all the properties of elementary particles withered away, as the less ambitious and more traditional quark-lepton theory met with success, after success to become today's Standard Model. Thus, Capra's argument that modern physics supports the holistic paradigm falls like a house of cards with the demise of S-matrix theory and the reaffirmation of atomicity in the Standard Model.

The Universe of Stephen Hawking

Although S-matrix theory is now out of fashion, the bootstrap idea has been revived in spirit, though not in name, by cosmologists who are working to develop a quantum theory of gravity. The best known of these new bootstrappers is Stephen Hawking, who occupies the Lucasian Chair of Mathematics of Cambridge University, once held by Isaac Newton.

In Hawking's best-selling popular book *A Brief History of Time* (1988), he relates his hope for a self-consistent theory of the universe that will admit only one solution: the universe as we perceive it. He proposes that this theory will be found within the mathematical framework of quantum gravitational theory.

Hawking argues that any successful quantum theory of gravity must treat time in an equivalent way to the three spatial coordinates. This requires that time be an imaginary number. If so, then all four spacetime dimensions may be boundless.

We can understand Hawking's picture if we think of a two-dimensional universe analogous to the two-dimensional surface of a sphere, with one space dimension given by the longitude and one time dimension given by the colatitude. That is, time is given by the angle measured from the North Pole to a circle of latitude.

In this picture, the universe is completely self-contained in time as well as space. Just as the spatial longitude coordinate on a sphere has no beginning or end (except an arbitrary meridian, such as Greenwich, that we define as zero), so, too, time has no beginning or end. With no boundaries, there can be no boundary conditions, including the familiar ones at time zero associated with the action of a creator. And with no boundary of time, there is no creation and no creator.

Einstein had once asked if God had any choice in creating the universe. According to Hawking, although God would have had no choice in terms of boundary conditions, he still might choose the laws the universe had to obey. But, Hawking adds, this may not be much of a choice, with only one or a small number of complete unified theories possible (Hawking 1988, p. 174).

Because of the attention given to Stephen Hawking and his widely read book, many laymen might be inclined to think that he is near to achieving the goal of a unified, self-consistent, theory. I am sure Hawking would be the first to protest that this is not the case, that what he has presented in *A Brief History of Time* is mostly a gleam in his eye, not a theory that is even close to making any kind of prediction that could be definitively tested by experiment or astronomical observation. Popular books on cosmology commonly speculate on the universe's origins.

As with S-matrix theory and other modern unification attempts such as the idea of superstrings, Hawking and his research colleagues are pursuing certain mathematical lines that look promising, but which more likely than not will someday be seen as a dead end. This is what happened with S-matrix theory and all other overambitious attempts at an ultimate theory of everything. In this business, failure is no disgrace. Even Einstein, the genius of the century, was not able to find his sought-after unified field theory.

I am by no means suggesting that we should stop trying to achieve the incredible goal of a theory of everything. This is a worthy occupation for brilliant people like Stephen Hawking. But my personal wager is that such a theory will never be found. The universe could just as well be the way it is because it happened that way, and we could be the way we are because the conditions were such that we developed, random step by random step, to where we are today.

Nothing forbids that the particular configuration of our universe could have happened by accident. Other, different universes could just as well have spontaneously appeared, and maybe did. An accidental origin to the universe is the most economical solution to the origin problem consistent with current knowledge, particularly the statistical interpretation of quantum mechanics. The Capra-Hawking view of a single self-consistent universe, with none other possible, is simply Newtonian determinism dressed up in fancy new clothes.

The Turning Point

In his enthusiasm for a new world view, Fritjof Capra has gone further off the deep end than is justified by the data. He has followed *The Tao of Physics* with another book, *The Turning Point* (Capra 1982). Here Capra provides answers for all the problems of the world within a framework of a dubious holistic view of the universe that he wrongly claims is supported by the new physics.

The tao of physics merges with the new consciousness and then evolves into the New Age, and Capra uncritically embraces almost every unsupported claim associated with that movement, from psychic phenomena to homeopathy and orgone energy.

Capra accepts the superluminal interpretation of the experiments that test Bell's theorem, ignoring the fact that other, more economical interpretations exist. He understands that Einsteinian locality is thus violated, but argues that, while the variables of classical physics are local, "those in quantum physics are nonlocal, they are instantaneous connections to the universe as a whole" (Capra 1982, p. 82).

But, as we have seen, quantum mechanics is perfectly local, since it does not violate Einstein's theory of relativity. Only nonobservable mathematical inventions, like the wavefunction or probability, are nonlocal.

Capra tries to avoid the conflict of his holistic ideas with relativity by arguing that the superluminal connections are not signals in the Einsteinian sense. "They transcend our notions of information transfer," he says (p. 85). Also, they are not "causal in the classical sense" (p. 89). But if they cannot transfer information, or physical quantities, such as energy and momentum, from one point to another, what do they transfer? How can events at one point influence events at another without being a signal?

Here again we can see that New Age physics is more consistent with the physics of the last century than of this century. If the aether did exist, then superluminal motion would be possible—just as supersonic motion occurs in the earth's atmosphere.

Capra ignores the success of the Standard Model of quarks and leptons, and the failure of S-matrix theory, claiming that modern physics shows "we cannot decompose the world into independently existing smallest units" (p. 81). But that's precisely what the Standard Model does, and it works, at least at the level of quarks and leptons. Capra's statement is simply false.

The bootstrap approach brings human consciousness into the picture as part of the self-consistency needed to complete the story (p. 95). Capra adopts the common fallacy that the quantum mechanical interaction of the measuring apparatus with the object being measured implies the intervention of human consciousness. Once again, we see the effect of human self-centeredness. We can't seem to shake ourselves of the arrogant notion, as expressed by Protagoras, that "man is the measure of all things."

The Systems View

Capra calls his holistic view of modern physics a "systems view." He says his systems view replaces the Newtonian mechanistic view that dominates modern thought, ignoring the fact that quantum mechanics is over sixty years old and physicists have not been thinking in Newtonian mechanistic terms for that long. In fact, Capra's view represents more of a return to mechanistic ideas than modern physics, with his invisible holistic fields controlling everything that happens. Capra's view is the "old physics" of the nineteenth century, the physics of Oliver Lodge and the aether.

Capra says that severe limitations exist in the conventional mechanistic molecular approach to biology (p. 121), and notes related deficiencies in what he calls a Newtonian approach to human behavior in psychology, and in economics, as manifested in worldwide economic crises.

These are all said to result from reductionism. The devil of reductionism is to be overthrown by the "new vision of reality . . . based on the awareness of the essential interrelatedness and interdependence of all phenomena—physical, biological, psychological, social, and cultural" (p. 265). Our physical, mental, social, and economic health will all be restored if only we apply the new systems techniques.

In the systems view, the world is composed not of quarks and leptons, or cells and organs, but of relationships that form an integrated whole. Reductionism is not totally discarded. That would be folly, in view of its great success. But holism, analysis, and synthesis are presented as complementary approaches to be put in "proper balance" (p. 268).

Self-Organizing Systems

Capra tells us that living organisms, in particular, are not machines, because "machines are constructed, whereas organisms grow" (p. 268). But crystals grow too. Oops, I forgot. This is the New Age, and crystals are also living "organisms" in that perspective.

Living organisms are certainly self-organizing systems. But Capra claims that their order and structure are imposed from within the system itself. That is, by virtue of their organization, they develop a certain autonomy. Life and all other organized structures have a will of their own: ". . . evolution is not dominated by 'blind chance' but represents an unfolding of order and complexity that can be seen as a learning process, involving autonomy and freedom of choice" (p. 288).

The systems view is extended to social and cultural structures; these are said to process information, learn, and have memory in a way that we associate with mind (p. 290). Mind and matter are not separate categories, in the Cartesian way, but different aspects of the same universal process (p. 290).

The Germ of an Idea

Actually, the systems view contains the germ of an idea having some merit, although the idea did not originate with Capra. Complex systems seem to be able to develop qualities of the type we normally associate with mind or consciousness. I will get back to this profound notion in the next chapter. But first, let me polish off Capra.

Capra tries to relate the self-organization of matter to mysticism, saying: "This relative concept of free will seems to be consistent with the view of mystical traditions that exhort their followers to transcend the notion of an isolated self and become aware that we are inseparable parts of the cosmos in which we are embedded" (p. 270). This is yet another connection I find rather strained.

Capra refers to the work of Nobel laureate Ilya Prigogine on the dynamics of self-organizing systems. According to Prigogine, these systems maintain and develop order by breaking down the order of other systems, thereby "creating entropy." Capra makes it sound like some kind of miracle— creating entropy—like creating life, or the universe.

But I must interject that nothing here involves profound new principles. Prigogine's entropy-creating systems do not violate any laws of physics. Organization takes place in nonequilibrium systems. Nineteenth-century thermodynamicists knew that, and I am sure that Prigogine was not suggesting that there is anything paranormal about entropy production. Want to create entropy? Just rub your hands together.

The imagined conflict between the systems view and reductionism was already referred to in Chapter 6. We saw that similar ideas were expressed in that context by the Nobel laureate Roger Sperry. Sperry said that power flows in both directions: upward from the particles and forces that make up matter, but also downward from the whole structure itself. In this view, the organized whole has a life of its own—like a living organism whose atoms and molecules continually replace themselves while the body retains its identity in their structure and organization.

This view also finds application in the "Gaia" hypothesis of chemist James Lovelock: the earth itself is like a living organism, a living being (Lovelock 1979). As we will see, when stripped of its mysticism and holistic miracles, the natural, spontaneous evolution of material systems to higher levels of complexity provides precisely the model we need to understand the development of life and consciousness without the intercession of transcendent forces.

The Cosmic Mind

Capra then takes a huge leap, concluding that all these "minds" that are produced as complex systems develop are really one. "In the stratified order of nature, individual human minds are embedded in the larger minds of social and ecological systems, and these are integrated into the planetary mental system—the mind of Gaia—which in turn must participate in some kind of universal or cosmic mind." This sounds a little like Spinoza's God. Capra's God is "the self-organizing dynamics of the entire cosmos" (p. 292).

People are a part of this great cosmic whole, and societies and cultures all possess a common consciousness (p. 296). Again, Capra sees similar tenets in various mystical traditions: a nonmaterial, formless "pure consciousness," "ultimate reality," "suchness," and the like (p. 297).

Of course, the human component of the cosmic mind cannot be the one provided by reductionist neuroscience. Instead, Capra urges something holistic, perhaps along the lines of the holographic model of the brain of neuropsychologist Karl Pribram (Pribram 1977).

Holograms

The hologram has provided a convenient model for concepts that connect the whole and its parts. In an ordinary photograph, a point-to-point correspondence exists between the film and the image. By contrast, each small section of a hologram contains all the information to reproduce an entire picture, although the smaller the section used, the less sharp the result. Furthermore, information about each spot on the image is contained in the entire hologram.

At the time of Pribram's proposal of the holographic brain, visual memory had not been isolated within the brain, so he speculated that it was perhaps like a hologram. A clever idea, but progress in brain research since then has supported the more conventional reductionist view of a brain with largely separated functions. New devices, such as the PET scanner discussed earlier, have highlighted localized regions of the brain that seem to be responsible for the various aspects of visual perception as well as many other types of mental activity.

Bohm's Enfolded Order

The remarkable properties of the hologram have also inspired physicist David Bohm to speculate on an analogous holistic model of the universe (see the discussion in Zukav 1979). Recall that it was Bohm who first proposed the type of experimental test of the EPR paradox that was eventually implemented. He has been the leader of the small anti-Copenhagen group of physicists looking for an alternative hidden variables interpretation of quantum mechanics. Recognizing that local hidden variables are now ruled out by the experimental test of Bell's inequality, Bohm has sought his goal within the framework of nonlocal, or holistic, hidden variables.

Bohm suggests we turn physics around, starting with the whole and working down to the parts. The universe we see may appear to be composed of discontiguous particles, which Bohm called the "explicate order," but this results from an "unfolding" of an unbroken whole at the more fundamental level called the "implicate order."

In terms of the hologram analogy, the explicate order is the normal photographic image, formed in one-to-one correspondence to the light we see from a three-dimensional object. The implicate order is the hologram, where the information for each point on the image is contained on the entire surface. In Bohm's analogy, we live in and observe the particulate unfolded order that is the photographic image, but the hologram contains the enfolded order that is the true reality of the universe.

Reducing the Question

So what are we to make of these many fine-sounding ideas? Being a reductionist, I can only come to grips with them by reducing them to their essentials. I am sure holists will criticize me for this approach, as well as for my conclusions.

Basically the holistic ideas surveyed above come down to the view that the universe is an unbroken wholeness that can only be fully understood by a consideration of the whole. Reductionism holds that the universe consists of parts that interact with one another only locally—meaning nonsimultaneously. The holist says that this is an approximation that works sometimes, but not all the time, and especially not if we attempt to understand the essence of the universe, which can only be done holistically.

If the universe is an unbroken whole, then certain interactions between what are normally assumed to be separate parts, like the minds of individual human beings, can take place. We are incorrect when we insist on thinking only in terms of discrete parts.

Now if we don't think in terms of parts, what do we think in terms of? Actually, holists still divide the world up into various parts, like Gaia earth and cosmic mind. So I feel free to continue to do so too; this seems to be a natural and necessary way for us to classify our sense observations. Until Bohm or someone else tells me explicitly how to deal with the implicate order, I can only make implications within the framework of the explicate order.

So the question then is reduced (that word again!) to whether the parts into which we conventionally divide the universe act sufficiently independently that we can focus on them individually without worrying about all the other parts at the same time. Alternatively, the simultaneous interactions of all the parts must be considered. How do we distinguish between these two possibilities? The usual way: empirical test.

An important clarification must be made at this point. We have to distinguish between the two forms of holistic conceptions, which I have labeled "strong" and "weak." The weak form of holism merely says that we can find principles that describe the gross behavior of complex systems, principles that bear little or no resemblance to the reductionist laws obeyed by the parts of the system. These could range from rules of organic evolution to principles of human behavior to ideas of esthetics. We don't derive them from the Standard Model of elementary particles and forces. But they don't violate the Standard Model, either.

If weak holism means that a cathedral is more than a pile of rocks, or that medical doctors should treat their patients as complete beings, I have no quibble with that. Weak holism is a rather obvious and noncontroversial concept. Strong holism, on the other hand, with its claims of miraculous power, raises issues of a far more fundamental nature.

Strong Holism

Strong holistic principles are more than simple-minded homilies about wholeness. By no means do they constitute a natural next step in the development of twentieth-century science, as is frequently claimed. Rather, they strike at its very heart, overthrowing the discreteness, locality, and even indeterminacy of the new physics.

In the strong form of holism, such great force flows from the whole to the parts that miracles happen. All the principles of microscopic physics— Einsteinian locality and quantum discreteness, uncertainty, and indeterminacy—are destroyed by this power of the whole.

Locality is violated by the notion of instantaneous connections between events. To argue that these superluminal connections are not the "signals" forbidden by Einstein's relativity begs the question. If no information is transferred, then no observable effects can occur. To say a connection exists without information transfer is like saying that entities exist that do not interact with the universe, and so have no effect on anything. It is an empty, useless statement.

Another logical inconsistency of New Age holistic thinking is that quantum mechanics provides a scientific basis for departing from the atomic view of nature. Quite the contrary. To quantize something is to limit it to a set of discrete values. Quantum mechanics arose out of the discoveries of the discrete nature of matter and light.

The core principle of quantum mechanics is that the universe is made of discrete parts. To somehow turn this around and say that quantum mechanics supports the notion of a continuous whole is an incredible feat of misrepresentation.

A similar misrepresentation occurs with the statement, by Capra and many others, that conventional science has given us a machine universe in which humans have no free will and act only in ways required by the deterministic laws of the machine. Again, the opposite is true. Quantum physics, with its Heisenberg Uncertainty Principle, did away with the Newtonian World Machine. Holistic thinkers want to put it back, with deterministic hidden variables replacing the statistical wavefunctions of the Copenhagen Interpretation.

The idea that everything happens according to fate, because "it is written" in the grand plan of a transcendent spiritual reality, is the ancient teaching of most religions, not modern science. Far from limiting human free will, modern reductionist, relativistic, quantum science is the only current human system of ideas that permits it!

Searching for Evidence

So I claim that current conventional science provides no support for the beliefs of the traditional churches or the secular spiritualism whose current incarnation is the New Age. Rather, I see only an irreconcilable conflict of basic assumptions.

How can we test strong holism? The same way we test any paranormal or supernatural claim: look for a miracle; find a convincing demonstration of the breakdown of reductionist science.

Unfortunately, most of the proclaimed miracles attributed to holistic forces are the types of cures or psychic experiences that have occupied us in previous pages. Moreover, they involve humans, who, for a variety of reasons, have a tendency to mistake the true causes of events. We have seen how difficult it is to rule out normal explanations for miraculous claims, including fraud, scientific incompetence, and self-delusion. But such explanations cannot be ignored simply because they are unpleasant to consider.

No experiment done with human subjects has ever provided conclusive evidence for any phenomenon violating the hypotheses of reductionist, materialist science, despite countless claims and testimonials. Neither has the sum total of all experiments and anecdotal reports provided such evidence. The experiments and reports provide us with insufficient reason to consider the additional hypothesis of strong holism.

Someday it may be possible to devise flawless experiments involving humans, but more reliable tests are possible if we minimize the role humans play in the experiments. By implication, it may be futile to try to test for psychic phenomena associated with human minds. But if mind is an emergent property of any system that attains a sufficient level of organization, which even materialist science allows is possible, and if it can be demonstrated that such a property has the power to violate the laws of microscopic physics, then we should be able to find evidence of such power in nonhuman systems. Having found it, a strong case could then be made that the human mind also has this power.

The ideal test would be without human subjects in a well-equipped, immaculately clean laboratory containing a number of highly trained and skillful researchers. The researchers should follow very careful double-blind protocols, and present all the details of their procedures and data, so they could be critically reviewed and independently replicated. Is this possible? Of course. Thousands of such laboratories exist around the world, and turn out example after example of solid, dependable, reductionist science. All we need is for a few of them to turn their attention to testing the claims of holistic science.

Water that Remembers

One laboratory that claims to have uncovered a violation of reductionist science is INSERM 200, the unit of immunopharmacology and allergy of the French National Institute of Health and Medical Research, located in the suburbs of Paris. In June 1988, the director of the laboratory, Dr. Jacques Benveniste, and twelve collaborators, published an article in *Nature* claiming they had found evidence for holistic effects irreducible to the known behavior of atoms and molecules.

The experiment involved exposing a type of white blood cell (human polymorphonuclear basophils) to certain antibodies (anti-IgE) that cause the blood cells to release histamine and increase their staining properties, allowing the effect to be measured. Diluting the antibodies in water to the point where the probability of finding even one molecule in the water was as low as 10^{-106}, antibody action was claimed when the solution was "accompanied by vigorous shaking" (Davenas 1988).

How was it possible that water could maintain a property of an antibody solution without any antibodies present in the solution? Benveniste and his coauthors say that "specific information must have been transmitted during the dilution/shaking process." As for a mechanism, they suggest that the structure of the water was changed when the antibodies were present, and that this structure was maintained and passed on as more water was added and mixed in with the original.

If these results are correct, somehow a system of simple molecules of liquid H_2O was able to remember its previous properties. Benveniste was quoted in the August 8, 1988, issue of *Time* magazine: "It's like agitating a car key in the river, going miles downstream, extracting a few drops of water, and then starting one's car with the water."

The suggestion that water could remember recalls the emergent holistic properties we have heard proposed. If the Benveniste laboratory results could be confirmed and replicated, we might have to discard some of our most sacred reductionist ideas. We would be forced to look seriously for alternative models that would make it possible for collections of atoms as simple as water to develop mind-like capabilities.

Homeopathy

If correct, the INSERM 200 results tended to confirm the principles of the fringe healing method called homeopathy that is widely practiced in France. In fact, some of the laboratory's financial support came from French manufacturers of homeopathic nostrums.

The practice of homeopathy is traced back to India's ancient Ayurvedic medicine and Greece's Hippocrates. Certainly all cultures have learned, from trial-and-error, that various herbs and other plants can provide relief from many ailments. Modern medicine has made use of much of what has been learned in the experiences of the folk medicine of the ages.

But homeopathy is alleged to be much more. In the late eighteenth century, a German chemist and physician named Samuel Hahnemann developed the concepts that are the foundation of the practice today. The basic claim is that highly dilute portions of normally harmful substances, if vigorously shaken, can provide healing power.

The homeopathic "law of similars" states that the medicine used to treat a certain set of symptoms in a sick person will produce those symptoms in a healthy person (Anderson 1978).

On the surface, this sounds like the vaccines of conventional medicine, which work by stimulating the immune system to produce the specific antibodies that fight a particular disease. The problem with homeopathic remedies, however, is that the dilutions used are often so high that not a single molecule of the active ingredient should be present, based on what we know about atoms and molecules. In fact, the claim is made that the higher the dilution, the longer lasting the results (though the slower to take effect).

While treatments are slow, often taking many months, their efficacy, if any, cannot be explained by conventional biology, for example, as the simple accumulation of the active substance. Only a few molecules, if any, are likely to have been present in the individual treatments.

Homeopathic theory relies on the ancient idea that living beings have a "vital force" within them that is "the energy which animates and drives the human being and which integrates the mind, body and soul of man." The concept is likened to *prana* in yogic philosophy, the Chinese *ch'i,* Japanese *ki,* the electricity of a TV set, and "The Force" in the film *Star Wars* (Anderson 1978, Ullman 1988). Impressive credentials, to say the least.

The Dilution Delusion

The INSERM 200 paper in *Nature* was accompanied by an editorial disclaimer that the laboratory's results should be accepted with caution, and noting that "there is no objective explanation for the results."

Nature had issued a similar disclaimer when it published the Stanford Research Institute experiments with Uri Geller in 1974. This time, however, *Nature* went further. Perhaps because of the criticism encountered in the Geller case, editor John Maddox led a team to Paris to investigate the procedures of Benveniste's laboratory first hand. This week-long visit was a precondition for the publication of the original article.

Maddox took along Walter Stewart, who investigates fraud for the U.S. National Institutes of Health, and paranormal debunker James Randi.

Dr. Benveniste and the laboratory's employees were cooperative, demonstrating the experimental procedure and providing lab notebooks for photocopying. They also repeated the experiments with blind protocols provided by the investigators, who were expert in designing experiments to guard against fraud and the other typical ways that data can be contaminated.

None of the three experimental runs performed under the investigators' protocols reproduced the positive effects of antibody action reported originally, although four runs done in the presence of the *Nature* team by the laboratory's ordinary procedures did show the reported effects.

The *Nature* team's report, published in the July 28, 1988, issue, pulled no punches. It concluded that the experiments were poorly controlled and not free from systematic errors, including observer bias. The *Nature* report said that the extraordinary claims made and their interpretation were not matched by the care taken in performing the experiments.

As the saying goes, extraordinary claims require extraordinary proof. In fact, the INSERM 200 experimental techniques were poor even by ordinary standards used for ordinary claims. The positive results were not reproducible when the experiment was properly controlled, and further suffered from problems of statistical inconsistency (Maddox 1988).

The investigating team did not find any evidence for fraud. Maddox is quoted in the *Time* magazine article as saying: "We believe the laboratory has fostered and then cherished a delusion about the interpretation of its data."

Nature allowed Benveniste to respond in an article in the same issue. He attacked his critics as "a squad of 'self-appointed keepers of the scientific conscience.' " Further, he insisted that "this kind of inquiry must immediately be stopped throughout the world," since "Salem witchhunts or McCarthy-like prosecutions will kill science" (Benveniste 1988).

The reference to witchhunts and McCarthyism struck a familiar chord for me. Almost the identical words were used against my fellow skeptics and me in the Hawaii psychic affair. One sure mark of pseudoscience is that investigators whose work is criticized make personal attacks on their critics in place of answering the specific criticisms. I guess pseudoscientists think if they can convince the public that skeptics are terrible people, then their delusions will somehow become true.

References

Anderson, David, Buegel, Dale, and Chernin, Dennis. 1978. *Homeopathic Remedies for Physicians, Laymen and Therapists.* Honesdale, Pa.: Himalayan International Institute.

Ashvaghosha. 1900. *The Awakening of Faith.* Suzuki, D. T., trans. Chicago: Open Court.

Benveniste, Jacques. 1988. "Dr. Jacques Benveniste Replies." *Nature* 334:291.

Capra, Fritjof. 1975. *The Tao of Physics.* Berkeley, Calif.: Shambhala.

———. 1982. *The Turning Point.* New York: Simon and Schuster.

Ch'u Ta-Kao. 1973. Trans. of Lao Tsu, *Tao Te Ching.* New York: Vintage Books.

Davenas, E., et al. 1988. "Human Basophil Degranualation Triggered by Very Dilute Antiserum Against IgE." *Nature* 333:816.

Fung, Yu-Lan. 1958. *A Short History of Chinese Philosophy.* New York: Macmillan.

Hawking, Stephen W. 1988. *A Brief History of Time.* New York: Bantam.

Lodge, Oliver. 1914. *Continuity. The Presidential Address to the British Association for the Advancement of Science, 1913.* New York: Putnam. P. 21,23.

———. 1920. *Beyond Physics.* London: Alan and Unwin.

Lovelock, J. E. 1979. *Gaia.* Oxford: Oxford University Press.

Maddox, John, Randi, James, and Stewart, Walter W. 1988. " 'High-Dilution' Experiments a Delusion." *Nature* 334:287–290.

Oppenheim, Janet. May 1986. "Physics and Psychic Research in Victorian and Edwardian England." *Physics Today.* P. 62.

Pribram, Karl H. 1977. "Holonomy and Structure in the Organization of Perception." In *Images, Perception and Knowledge.* Nicholas, John M., ed. Dordrecht-Holland: Reidel.

Suzuki, D. T. 1952. *Studies in the Lankavatra Sutra.* London: Routledge & Kegan Paul.

Ullman, Dana. 1988. *Homeopathy: Medicine for the 21st Century.* Berkeley, Calif.: North Atlantic Books.

Zukav, Gary. 1979. *The Dancing Wu Li Masters.* New York: Morrow.

13.

Emergence

*Yet the existence of life is only one out of many complex and interesting arrange-
ments of molecules—one that suits the environment of earth. There could be
other equally complex and interesting combinations of molecules which could
be constructed artificially. No one knows the limits of what can be made from
atoms.*

Heinz R. Pagels

A Nagging Element

Reductionist materialism remains unscathed as the most workable model
of the universe of our senses. No phenomenon has been established that
violates any qualitative feature or quantitative prediction of the Standard
Model of physics. In particular, the qualities and phenomena that we associate
with mind or consciousness do not force us to the view that an additional
immaterial component must be added to the universe, other than the
immaterial words, symbols, and concepts we use to describe our experiences.

Yet, in all of this, mixed in among the chaos of unsupported occult
ideas of the New Age and the more orderly deliberations of highly competent
scientists such as Roger Sperry, a nagging element persists. While not
demonstrating any power to violate our reductionist scientific principles,
phenomena occur that do not seem to arise straight out of those principles.
These include most of the properties of many particle molecular and
biological systems, especially the lifelike and mindlike qualities that material
systems take on when they reach high levels of complexity.

Complexity

Scientists are still grappling with a precise definition of complexity. Clearly it cannot be trivially equated to number. Complex systems cannot be defined simply as having a large number of elements. A string of a million ones in a row has many elements, but it is very simple. The number of air molecules in a room is huge, yet the behavior of the air can be adequately described by a few parameters such as pressure, volume, and temperature.

Complexity refers to situations with high degrees of structure that cannot be described with a small number of parameters. The description of a complex system must involve many words, symbols, or binary bits of data. If the system is described in the language of bits, to which words and symbols can be reduced, the sequence of bits cannot follow a pattern reducible to a simple algorithm. If the pattern could be reduced to a short algorithm, the system would still be simple, since it could be replaced by the algorithm (Pagels 1989).

However, even when a structure must be described by a large sequence of bits, it may not possess the qualities we would associate with the notion of complexity. The difficulty comes in distinguishing a complex pattern from a completely random one. The string of bits that describes the placement of each brick in a magnificent building would look random to an observer who did not already know that they represented the positions of bricks. It seems you need to be able to see the results before you can decide on whether the pattern is complex or simply random, or at least be given more information than the bit sequence alone. Yet complexity is not equivalent to chaos. The idea of structure or form is embedded in the notion of complexity, whereas random systems lack structure.

Another reason we do not associate complexity with chaos is that highly random systems can be described simply with statistical mechanics. Using statistical mechanics, a random system can be described by a few parameters, such as pressure and temperature, and quite successful predictions of the average behavior of the system can be made. The motions of individual particles are unpredictable, but usually no one cares about these; when the number of particles is large, the fluctuations are correspondingly small, so that the calculated averages are adequate for most practical purposes.

Chaos Begets Order

So complexity lies somewhere between simple order and complete chaos. We are accustomed to observing simple order beget complex order, such as in the construction of a building. Construction workers lay out the bricks according to the architect's plans, using operations that can be described quite simply. Complex order clearly can be produced under the action of external ordering agents.

We are also familiar with many ways that simple order can lead to chaos, such as in an explosion or less violent mixing together of previously differentiated components.

Less familiar is the notion that chaos can beget order, spontaneously, without an external agent or plan. This is called self-organization. In a previous chapter I summarized the conclusion of my earlier book *Not by Design: The Origin of the Universe* (Stenger 1988), which tells how the universe itself could have begun in complete chaos, with no design or plan, as even the fundamental particles and forces arose by a process of spontaneous symmetry breaking. The universe, in other words, could very well be a self-organized structure. This model of the origin of the universe was suggested by examples of self-organization closer to everyday experience.

Recall the example of a snowflake, which is formed as water vapor freezes. The detailed pattern of that snowflake was not present in the original water vapor, stored in some yet-undiscovered memory system within H_2O. Rather, the snowflake pattern develops randomly, only restricted by certain atomic properties, to produce a six-pointed configuration for the snowflake structure. Randomness is superimposed upon order.

Other familiar examples of self-organization include rivers sculpting continents, a sperm and egg forming a human being, the speciation of life on earth, the self-assembly of a virus, and even the aging process in animals (Yates 1987, p. 2). In fact, virtually every complex biological and geological system on earth may be an example of self-organization.

The properties of such complex systems emerge spontaneously from the matter out of which they are composed. That is, the details of the structures are not inherent in either the original matter or in any of the laws of physics that are obeyed by the matter. These laws provide certain constraints, especially the limitations of growth imposed by energy conservation, but otherwise the emergent structures exhibit a large amount of autonomy. They almost seem to have a mind of their own.

Complexity and Consciousness

Pierre Teilhard de Chardin (1881–1955), the Jesuit paleontologist who was at one point barred by the Church in France from teaching evolution, equated material complexity with consciousness: "Whatever instance we may think of, we may be sure that every time a richer and better organized structure will correspond to the more developed consciousness." This he called the "law of consciousness and complexity" (Teilhard 1959). In his many writings,

Teilhard explored and developed the idea that life and consciousness evolve from matter as it moves to ever-increasing levels of complexity.

Today we are witnessing the rise of new "sciences of complexity" in which investigations are being made into systems of many connecting elements that heretofore have been too difficult to study with even the most sophisticated mathematical techniques. These new sciences include cellular automata, computer and information science, artificial intelligence, and chaos theory. Established sciences, such as genetics, neuroscience, meteorology, and many others, are also feeling the impact of the new sciences of complexity.

The remarkably powerful mathematical methods that have been applied over the years to few-particle systems, such as diatomic molecules, are of no use for systems of thousands of molecules, such as a strand of DNA. On the other hand, the methods of statistical mechanics, which work so well for systems composed of 10^{24} diatomic molecules randomly moving about in a container, are also of no use for more-structured, many-particle systems like DNA. Complex systems defy our finest mathematical physics.

What has brought the study of complexity into the realm of the possible is the computer. By means of computer simulations, complex systems or their analogs can be followed as they evolve with time, often with startling results.

Natural Evolution

Lifelike qualities can be understood to have evolved quite naturally and spontaneously, once a sufficiently complex molecule chanced to develop the ability to reproduce itself. In conjunction with the mutations that accidentally occurred from background radiation and other means, this led very rapidly to the development of self-replicating and self-organizing molecular systems of ever-increasing complexity.

The key is self-duplication. The great mathematician John von Neumann developed a logical proof that it is theoretically possible to build a machine capable of reproducing itself (Von Neumann 1966). Our DNA genetic structure is one particularly successful implementation of a Von Neumann machine.

Self-organization happens without any violation of physical principles such as the second law of thermodynamics. The second law does not forbid the spontaneous formation of more organized structures; it merely requires that a price be paid by increased disorganization elsewhere. Complex systems that are far from equilibrium and in constant interaction with their environment are able to utilize energy from that environment in their self-ordering processes. As a system becomes orderly, its own entropy must decrease; but this can happen as long as the entropy, or disorder, of the environment increases by an equal or greater amount.

The features we now associate with life appear *gradually* as evolving systems become more complex. I emphasize gradually, because at no point does it seem necessary to introduce a sharp dividing line between living and nonliving, where some spark of life enters the system. Richard Dawkins has argued that species evolve as the result of tiny changes, not the large quantum steps (punctuated evolution) that other evolutionary theorists have postulated (Dawkins 1987). But neither theory requires supernatural intervention, and punctuated evolution is not so punctuated that everything is compressed into six days.

Once particularly useful features emerge in the evolutionary process, they very rapidly become dominant, that is, when viewed on a time scale of millions of years. Thus, while that first replicating molecule undoubtedly did not utilize our current DNA-based reproduction scheme, once that scheme accidentally developed, it provided such improved survival value that other methods would have quickly died out.

Another randomly developed feature that provided greatly enhanced survival value was the nervous system. This enabled organisms to be active rather than passive in their quest for nourishment and in fulfilling their overwhelming imperative to have their genetic infomation passed on to the next generation. Eventually, and again gradually with no outside help, the nervous system evolved a brain. And with the development of the brain emerged the qualities we identify as mind and consciousness.

No Magical Moment

Ironically, the evolution of life and mind required randomness, in the mutations that led to the changes necessary for development. How contrary to the view that life and mind resulted from the design of a creator! How more akin is religious thinking to Newtonian determinism than to modern scientific thinking!

The mental properties that develop with increasing complexity and organization, like the properties associated with life, did not just happen at one magical moment with the creation of a first human being. There was no first human being, no Adam, and there was no first human consciousness.

We can draw the inference that mental abilities will emerge whenever any system of matter, whether it be made of cells of proteins and other organic molecules, silicon computer chips, or perhaps even dark matter somewhere deep in space, reaches a sufficient (though not sharply defined) level of complexity. Teilhard thought that some aspect of consciousness is present—though at a very low level—even in simple assemblages of atoms. The level of consciousness becomes progressively higher as systems become more complex.

And these systems are not limited to the groupings of atoms and molecules into some physical structure. Mindlike qualities can result from any system composed of many strongly interacting parts—like a computer program, an artificial neural network, a human culture, or an economic system.

Groups of people form complex societies that evolve mindlike qualities. Sometimes, they also seem to develop a mind of their own, defying attempts to control them. And perhaps the Gaia hypothesis, that the biosphere of earth acts as a living organism, has some merit—as long as we recognize that this does not imply any profound spiritual contact with a cosmic mind, but just that the earth constitutes a complex, highly interactive material system.

Computer Consciousness

The strongest argument against the belief that emergent mental abilities are immaterial is that purely material computers can be used to simulate, at least primitively, the evolution of such qualities. As far back as the 1960s, when computers were in their infancy, with far lesser capabilities than now, people were already beginning to write programs that illustrated the development of artificial intelligence by Darwinian natural selection (see, for example, Fogel 1966).

The emergence of complex structures from a few simple rules that do not explicitly contain the structures themselves has been known for many years. These structures are not limited to static patterns, but exhibit dynamical features that we can identify as lifelike and mindlike in nature. Starting out as amusing board games, the simulation of self-organization on the computer has now evolved into a serious study of complexity.

Implementing John von Neumann's mathematical theory of cellular automata, the computer screen becomes a universe in which dots of light trace out fantastic patterns that evolve, reproduce, and branch out into unexpected forms. All this happens with the simplest starting rules. (For some beautiful examples of the types of structures called "fractals," see Gleick 1987.)

Self-reproduction can be illustrated by a simple algorithm developed by Edward Fredkin in the 1960s (Gardner 1971, Poundstone 1985). Fredkin's game can be played on a checkerboard or graph paper, but it is much easier on a computer.

Think of the computer screen as a grid, like a large checkerboard. Starting with any initial pattern of dots on the grid, you rewrite the screen

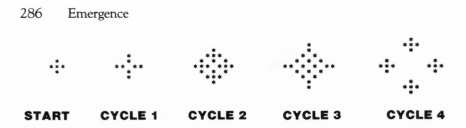

START   CYCLE 1   CYCLE 2   CYCLE 3   CYCLE 4

Figure 13.1. Example of a self-reproducing cellular automaton using Fredkin's algorithm. The initial pattern produces four copies of itself every four cycles. The same algorithm will cause any initial pattern to self-reproduce.

so that a grid cell, or pixel, is "on" if and only if an odd number of the four nondiagonal (above, below, left, right) neighbors of the cell is currently on. Repeating this process cyclically, the original pattern will reproduce four exact copies of itelf every n cycles, where n is a power of two. For example, in Figure 13.1 we see that a simple cross will reproduce four copies of itself in four cycles. This example is simple enough that you can produce it yourself, with graph paper and a little patience.

The Game of Life

Even more remarkable and unexpected behavior can be demonstrated with the program "Life" developed by John Horton Conway in 1970 (Gardner 1971, Berlekamp 1982, Poundstone 1985). The rules are almost as simple as in Fredkin's game, and are meant to simulate the types of forces that come into play in evolutionary processes.

Again, as in most cellular automata, a grid cell is turned on or off for the next cycle according to the current status of the cell's neighbors. In this case, we consider all eight neighbors of the cell, including those on the diagonals. A cell that is on is kept on for the next generation if two or three neighbors are on. It is turned off, however, if four or more neighbors are on, if one neighbor is on, or if no neighbors are on. A cell that is off is turned on if three neighbors are on. Thus, a cell needs company to be born and survive, but can die from overpopulation. In a crude way, this suggests the primal forces affecting living things; hence, Conway called his algorithm the Game of Life.

The structures that emerge from the most innocuous starting patterns amaze everyone who has played the Game of Life. These structures are totally unexpected—not programmed into the original algorithm or even dreamed of by the original programmer.

The structures of Life sometimes stabilize, but more often they change and evolve. Some, called "gliders," move across the screen. I have illustrated these in Figure 13.2. Again it is possible for the reader to reproduce this pattern using graph paper. Other figures oscillate between different forms,

Figure 13.2. The "glider" in Conway's game of *Life*. It reproduces every four cycles, moving across the screen to the right and downward.

or blink on and off. Collisions between moving structures can result in annihilation of the structures or new structures that evolve further or stabilize. "Glider-guns," that regularly shoot off gliders, have been discovered.

Today, with personal computers almost as common as TVs and VCRs, anyone can demonstrate the emergence of structure for herself with cellular automata games. I have done it at home on the same inexpensive home computer I am now using as a wordprocessor. Although I have access to more powerful machines, including supercomputers, through the terminal in my campus office, I didn't need them for the simple exercises of demonstrating the principles involved. I was able to type a slow version of Life into my home computer and get it running in under an hour. However, special programs and large computer screens are necessary to demonstrate the more profound features of the game.

Simulated Evolution

In *The Blind Watchmaker,* geneticist Richard Dawkins showed how the computer can be used to simulate the selection processes of evolution; his program generates buglike structures he calls "biomorphs" that evolve by "artificial selection" (Dawkins 1987). For a nominal fee, purchasers of his book are able to get a copy of the program, which runs on the Apple Macintosh computer, and have some fun.

Dawkins recognized that his biomorphs do not quite evolve naturally, so he issued a challenge to progammers everywhere to write a program that demonstrated natural evolution. This challenge was taken up by Michael Palmiter, a high school teacher in California, who developed a program called "Simulated Evolution" (copyrighted; available from Life Sciences Associates, One Fenimore Road, Bayport, NY 11705).

The main features of Palmiter's program were outlined in the Computer Recreations column of *Scientific American* (Dewdney 1989); I used Palmiter's ideas as outlined in the *Scientific American* article in writing my own program. I have also run Palmiter's superior original, graciously provided to me by Life Sciences Associates.

Palmiter's algorithm, as outlined in *Scientific American* and the materials provided by the distributer, is as follows: a unit called a "bug" can be found at any point on the computer screen. It moves in any one of six directions, in steps of one or two pixel units at a time. Which direction it moves in is determined by the values stored in a six-element array called the bug's "gene."

After a certain number of moves, the bug is able to replicate, producing two new bugs. These each have the same "genetic structure" as their parent, except for a random "mutation" in their genes. (Of course, living organisms do not mutate every generation, but this feature of the computer program speeds up the process.) As the bugs move across the screen in steps, they use up "energy," and they die when their energy dips to zero.

However, they are able to feed on energy located at various points on the screen, as they move about in what are initially random directions. That is, they start out stupid. But—and here's the point—succeeding generations of bugs get smarter and smarter, developing genetic traits that move them in directions toward greater food supplies.

For example, the tendency to step backward to where the food is already gone is quickly breed out of the tribe. Bug species that just move in a circle quickly die out as they devour all the food in their limited region. Likewise, bugs that move in one direction will die out, eventually colliding with the border of the screen and exhausting energy futilely pushing against it.

Other families of bugs, like the pioneers of the American West, naturally develop strategies that enable them to move off in directions of greater opportunity. For example, a species called "cruisers" tends to move forward about eighty percent of the time and backward only about one percent of the time. The remaining nineteen percent they move in some other direction. So they avoid going backward—a smart move, since the food behind them has already been eaten. But they also avoid the risky damn-the-torpedoes-full-speed-ahead strategy that would put them in danger of extinction. They follow a mostly forward but broken path that optimizes their survival.

It's a lot of fun to watch these qualities develop on the screen. The progeny of smart bugs sometimes get some dumb genes by mutation, and these die off, while progeny with smart genes continues on.

Evolution of Intelligence

And so intelligence evolves because of its survival value. Palmiter's bugs demonstrate, albeit simply and primitively, the basic qualities that we associate with life and mind. By using knowledge stored in their genes, they move and feed and reproduce, and they evolve into more intelligent species with enhanced survival capabilities. Is the difference between these bugs and humans any more than a quantitative one, made possible by the far greater storage capacity of our DNA? Those who think so will be hard-pressed to show me why.

It may seem that I am making a huge leap from a simple computer game to the evolution of intelligence. Obviously, the human nervous system is far more complex than the genes of Palmiter's bugs. Of course, no one has made a machine that exactly duplicates the processes in the human brain. But I claim that no one has to do so in order to prove my point. It suffices to find examples that demonstrate that the qualities we associate with intelligence can arise spontaneously in material systems. Once those examples are demonstrated, it becomes difficult to argue that material systems cannot, by self-organization, produce higher intelligence. Why should an external agent be required only above some unknown level of intelligence?

Intelligence is the ability to respond to stimuli in a way that provides the organism with optimal benefit. For most organisms, the goal is simply survival—so that the information stored in the individual's genes be passed to the next generation. The bugs in Simulated Evolution exhibit just this ability. Succeeding generations get smarter, better able to cope with their limited and competitive environment.

Even the simplest of real-life organisms are more complex than Palmiter's bugs and in the struggle to survive must deal with more varied stimuli. But only in a matter of degree. We humans have developed more sophisticated goals than simple survival. The need to survive remains primary, however, and has greater immediacy in many parts of the world. This, too, is just a matter of degree.

Selection processes obviously played a role in the development of the types of human intelligence that improve chances of survival. I have already noted that this did not evidently include a strong proclivity for rational thinking. Toolmaking ability, and social cooperation—both also intelligent qualities—have proved more useful for the greatest portion of humankind's evolution.

Evolving Complexity

So I am willing to go this far at most: I agree that, within the framework of the emergence of lifelike and mindlike properties from matter of evolving complexity, we may be able to discover profound new insights into the nature of our universe, and ourselves, that are not derivable within the strict reductionist framework of elementary particles and forces.

People who study properties of complex systems have to discover their operating principles by direct observation, not by calculations from the Standard Model. Chemists and biologists have their own laws, developed from observations in their own laboratories. Clearly these laws work. Perhaps chemists and biologists do not need to know about quarks and leptons after all, although you would think they still should like to know what was ultimately inside the systems to which they devote their lives.

The commonly heard argument that basic physics and chemistry cannot predict complex structures like specific enzymes may be true; but that same basic physics and chemistry predict that *some* kinds of structures will form under appropriate environmental conditions. And once structures have the ability to replicate, which can occur accidentally, they will naturally evolve into ever more complex and intelligent systems, better able to thrive in their environment and pass their genetic information on to the next generation.

The mistake made by those who argue that human life and mind are too unlikely to have developed by chance is the common one of egocentricity. We humans think we are the only possible form of intelligence. But as our computer games illustrate, self-organization and self-reproduction will occur in any sufficiently complex system. Our particular form of life was indeed unlikely. But some form of life on some planet was probably inevitable.

If we can't predict from the equations of physics what exact structures will develop, that's because, like next month's weather and the motion of electrons, the details are fundamentally unpredictable. They develop, like the universe itself developed ten or twenty billion years ago—with a strong component of chance. But as long as systems exist with numerous interacting elements, and these systems are far from being in equilibrium with their environment and have a nearby source of energy, structures will spontaneously form, and these will be expected to exhibit a strong natural desire to survive. This is our case on earth, where we are not in equilibrium with the much hotter sun.

This argument, if valid, would seem to suggest a high likelihood of life in the vicinity of many, if not most, stars in the cosmos. Some scientists have argued that this cannot be the case, because, if it were, some extraterrestrial beings would have contacted us by now. But this is again based on the egocentric notion that lifeforms will be something like us, that they will exhibit a similar desire to extend themselves to the heavens. Perhaps we are rare in this regard, and rare in the particular toolmaking abilities that make this possible.

The probability of earthlike life anywhere else but on earth is vanishingly small. Extraterrestrial lifeforms are probably so unlike us that any conclusions we draw based on our limited experience of a single planetary lifeform are highly problematical.

The Omega Point

Although the structure and complexity of the brain reached the level where human consciousness developed on earth, this development provides no support for the notion that these structures possess miraculous powers transcending the material world and raising them to some higher plane of existence. For, as we have seen, no one has yet been able to demonstrate that such powers exist.

If the human brain possesses no transcendent power, perhaps that will not be the case for other systems, on earth or beyond, that may develop even higher levels of complexity and consciousness. Possibly, evolution will lead some lifeform somewhere in the universe to a point where the level of mental ability will finally transcend the matter out of which it arose.

Despite our high opinion of the human species, and our desire to believe that we are the apex of evolution, I am afraid that we are probably not destined to be that super lifeform—at least until we have evolved for a few more millions or more likely billions of years. But don't despair. There's plenty of time left before all the stars in our universe burn out. And even if the human species does not itself evolve, perhaps our computer/cognizer progeny will.

The ultimate evolution to higher levels of consciousness was Teilhard de Chardin's vision, though he apparently still held a view in line with traditional Catholic teaching that humanity has already evolved to a pinnacle one step below divinity. If we discard that hubris, the primary ingredient remains: the march to higher complexity and consciousness continues, and ultimately converges on a structure of immense power and unity, the "Omega Point," that is the final crowning glory of evolution. Teilhard called this process "Christogenesis." At the Omega Point, matter becomes God.

References

Berlekamp, Elwyn, Conway, John, and Guy, Richard. 1982. *Winning Ways,* vol. 2, Chapter 25. New York: Academic Press.

Dawkins, Richard. 1987. *The Blind Watchmaker: Why the Evidence of Evolution Reveals a Universe Without Design.* New York: W. W. Norton.

Dewdney, A. K. May 1989. "Computer Recreations." *Scientific American.* P. 138.

Fogel, Lawrence J., Owens, Alvin J., and Walsh, Michael J. 1966. *Artificial Intelligence Through Simulated Evolution.* New York: Wiley.

Gardner, M. 1971. "On Cellular Automata, Self-Reproduction, the Garden of Eden and the Game of 'Life.' " *Scientific American* 224(2):112-117.

Gleick, James. 1987. *Chaos: Making a New Science.* New York: Penguin.

Pagels, Heinz R. 1989. *The Dreams of Reason: The Computer and the Rise of the Sciences of Complexity.* New York: Simon & Schuster.

Poundstone, William. 1985. *The Recursive Universe.* New York: Morrow.

Stenger, Victor J. 1988. *Not By Design: The Origin of the Universe.* Buffalo, N.Y.: Prometheus Books.

Teilhard de Chardin, Pierre. 1959. *The Phenomenon of Man.* Wall, Bernard, trans. New York: Harpers.

Von Neumann, John. 1966. *Theory of Self-Reproducing Automata.* Urbana, Ill.: University of Illinois Press.

Yates, F. Eugene, ed. 1987. *Self-Organizing Systems: The Emergence of Order.* New York: Plenum Press.

14.

emoving the Yoke

nexistent look much alike.

Delos B. McKown

Threads of the ꜱearch

And so we have found that the search for a world beyond the senses un-
ravels into many threads. Undoubtedly, I have not tied them all up to
everyone's satisfaction. First, we have had to deal with problems of definitions. What do we mean by natural and supernatural, normal and
paranormal, mind and consciousness? What is the difference between science and pseudoscience, or science and religion? I have tried not to be
too pedantic about these issues, taking the view that words mean what
most people take them to mean, and trying to judge what may be the
empirical content of the concepts that consensus attributes to those words.

Most agree that religion deals with the supernatural and science with
the natural, and that supernatural phenomena are basically miraculous,
breaking the laws of nature as we have come to understand them to be.

Religion has traditionally played several roles in society that are largely
independent of particular theologies or philosophies. They provide comfort
to the ill or grieving, and ceremonial rites of passage. Many cultural and
national rituals are carried on in a religious context. In some cases, as with
Judaism and Shintoism, a deep connection with ancient tribal character
is manifested by religious rites. People often participate in these rites out
of respect for their ancestors and with pride in their own national or ethnic

identity, without necessarily holding to the archaic supernatural beliefs associated with the actual rituals performed. That is not to say that they simply go through the motions. Surely deep feelings come into play, but they should not be misinterpreted as arising from any strict adherence to irrational ancient beliefs.

But the rituals and traditions of the religions of the world—rich as they are in artistic and cultural terms—are incidental to the fundamental view the religions profess about ultimate reality. All religions are founded on the principle that transcendent powers exist beyond those revealed to us by our senses, and that these powers play an abiding role in human life. Every church has leaders, ancient and modern, who claim special connections to those powers. This forms the basis of their claim of authority over people's actions.

The notion of religious authority deeply conflicts with the humanistic ideals of individual freedom and democracy upon which the United States of America was founded during the Age of Reason. Though often wrapped in the flag, and pandered to by hypocritical politicians, religious authority violates the spirit of the U.S. Constitution and the principle of separation of Church and State.

Religious authority also deeply conflicts with the message of science, which admits only the authority of observation and reason. Despite attempts to paper over these differences, science and religion are inherently incompatible.

Science Has Not Changed

The debate between science and religion, or science and supernaturalism, reduces in large measure to a debate on the existence of the miraculous. In the rhetoric of that debate we find gross misunderstandings of the nature of science. People argue that science is provisional, and so we can never know what the laws of nature truly are. Examples are given from the history of science that are meant to demonstrate how science has progressed by a sequence of revolutions, with old theories refuted and replaced by new ones that bear little resemblance to their predecessors. Since this happens in science, it is argued, then anything is possible. So why not miracles?

Well, a look at the history and practice of science reveals quite a different picture. Science began thousands of years ago, and although the volume of knowledge has expanded enormously in that time, the nature and methods of science have changed little. What is often interpreted as a great paradigm shift can be more accurately described as a clarification or reformulation of principles that were previously dimly perceived.

For example, Darwinism revolutionized the masses' understanding of humankind's place in the scheme of things. Yet evolution by natural selection had earlier been considered by others, notably Jean Baptiste de Lamarck (1744–1829). As much as 2,400 years earlier, the ancient Greek materialist Empedocles had suggested that the characteristics of species change with time. Still, at the time of Darwin, most people held that the human form was unchanged since creation.

Darwin recognized the deep significance of evolution, developed its theoretical framework, amassed a wealth of data to support it, and effectively communicated his results to the rest of the world. A tremendous achievement? Undoubtedly. Revolutionary in terms of the way humanity views its own origins? Certainly. But revolutionary in terms of the way science looks at the world? Not at all.

In developing the theory of evolution, Darwin applied conventional scientific method. His argument rested on the rational analysis of empirical data. The success of science in other areas had earned it respect and recognition, to the point where people began to understand the superiority of observation over human authority. The public's eventual acceptance of Darwin's theory arose out of its willingness to submit to the verdict of scientific method.

Similar misunderstanding of the history of science occurs in the common interpretation of the twentieth-century revolutions in physics. Einstein did not so much replace Newton as expand the magnificent structure of Newtonian mechanics. The relativity of space and time, the equivalence of mass and energy, were profound new ideas. But Einstein still measured time the same way Newton and Galileo did—with a clock. He used Newton's definitions of mass as the quantity of inertia, and momentum as the quantity of motion.

No basic physical quantity was introduced in relativistic physics that did not already exist, defined in the same operational way, in prerelativistic physics. Space and time remain the primary variables by which we represent events. Modern attempts to eliminate space and time, as with the S-matrix theory of elementary particles, have so far failed. Great physicists from Galileo to Einstein have clarified the meanings of space and time for us, not overthrown their basic conceptions nor declared them obsolete.

The laws of physics, as they were precisely and quantitatively formulated by Galileo, Newton, and their successors, have changed little in the centuries since their original development. Newton's three laws of motion remain unchanged to the present day, and continue to be applied in every aspect of modern technology. We still use Newton's second law, $F = ma$, to build the machines and structures of our technological world. We have flown to the moon with Newton's law of gravity, $F = Gm_1m_2/r^2$. The great conservation principles of momentum, energy, and angular momentum still apply, only now they are more deeply understood as statements about the basic symmetries of space and time.

Some principles of physics, such as mirror symmetry, have been found to be broken at very low levels and under special circumstances, but this breaking is now understood as a natural process. In fact, as we have seen, spontaneous symmetry breaking leads to the formation of structure and provides an important ingredient in the self-organization of material systems. Thus, symmetry principles are not invalidated by the discovery of tiny violations, but found to have even deeper meaning than we previously understood.

The instruments of modern science have provided us with greatly enhanced capabilities for gathering data about the universe. With our microscopes, telescopes, and particle detectors, we are no longer bound by the limitations of human sensory apparatus or of our confinement to this tiny planet. And we have learned to rely more on the rational interpretation of the reading of these instruments than on preconceived notions based on everyday experience.

Chasing Shadows

Those who claim an opening for the paranormal within the rapid change of modern science are like caged animals chasing moving shadows. The more science changes, the more it stays the same. We are free to call relativity a revolution. But what we shouldn't say, as many do, is that Einstein proved Newton "wrong." Within the bounds of the phenomena that Newton dealt with, he is as correct today as he was 300 years ago.

All religions teach that invisible transcendent forces lie behind all events in the universe. In the religions that are practiced by the greatest number of people, Christianity and Islam, those forces are controlled by a personal, humanlike God. By contrast, science is based on the assumption that events are products of detectable natural processes. Humanity is special only subjectively, because we ourselves are humans.

The notion of impersonal, natural processes determining events goes back to Thales. Heisenberg, Bohr, and their fellow quantum physicists discovered that determinism applies only to the large assemblies of atoms making up the objects of everyday experience, not to individual particles. Not every event requires a cause. We have learned that chance in nature is an equal partner with determinism in the description of events.

Not every event is predestined to occur by a set of preceding events. Some just happen. We are not automatons, controlled by what "is written." The universe itself could have been an accident. Still, the spirit of Thales is retained; the processes of nature are explicable without the intervention of undetectable transcendent forces.

The Most Successful Activity

We still maintain the atomic view of matter proposed by the ancient Greeks who invented science. In the intervening millennia, we have not changed their basic intuition of the discreteness of matter, just corrected the details. The main advice given to us by the materialist sages was to rely on our senses for knowledge about the universe, discarding all other claims to authority. We still follow this simple rule in science. Science is such a successful activity precisely because it deals only with what can be rationally defined and objectively measured.

When the Catholic Church wanted to determine the age of the Shroud of Turin, they turned to science for the answer, not Holy Scripture. When the pope gets ill, he says a prayer and calls in the papal physician. When Pat Robertson wants to get his message to the millions, he does not transmit it by ESP or some other form of spiritual communication; he buys time on TV. And he rides a limousine or helicopter to the TV station, not a flying carpet.

Most humans on this planet use the fruits of science in every phase of their lives. I become very irritated at those who decry science while accepting its every benefit. It is especially ironic how the antiscientists use modern communications to get their messages to the puiblic.

I'll Know the Supernatural When I See It

So does the supernatural exist? Let's look for ourselves and see. It's not difficult to think of events that could be classified as supernatural with a reasonably high degree of confidence. They simply have to violate those principles of physics confirmed and elaborated upon by centuries of careful observation. If we can show that some solid physics principle is violated in ways that cannot be understood as extensions beyond its realm of applicability, we will have made a good start toward demonstrating the existence of the miraculous.

Take the principle of energy conservation. Since energy is matter, if we can prove that matter has appeared sometime and someplace without a sufficient energy source to produce it, we will have demonstrated a miracle.

For example, suppose an object suddenly materializes in an empty field. It does not have to be an angel or the Virgin Mary; something prosaic like a large rock will suffice. The mc^2 of a woman or a rock is far greater than the energies of the photons from the sun striking the field and the thermal energies of the surrounding air particles. So, such an appearance would be evidence for a violation of energy conservation—a miracle.

Or, take the second law of thermodynamics. It forbids the spontaneous organization of a multiparticle physical system without a corresponding increase in the disorganization of the environment. For example, a drop of water fully insulated from the outside will not freeze into a crystal of ice. Heat must be removed, which then disorders the environment as much as, or more than, the crystallization process organizes the water molecules. Suppose we observed water convert to ice inside an isolated container or in a hot oven. Then we would have a miracle.

So, I see no great problem in identifying supernatural events when they are sufficiently obvious. While it may be difficult to arrive at a completely foolproof definition of a miracle that would satisfy a philosopher or a lawyer, I think the supernatural is a bit like pornography: I'll know it when I see it.

You would think, if the supernatural really plays the powerful and pervasive role in our lives so many people claim, its existence would have been obvious to all of us by now, as obvious as the fact that the sun lights the sky during the day.

But we cannot so quickly rule out the less than obvious. Is there still some extremely low-level phenomenon that we have simply not yet been able to verify? Leaving the philosophers and lawyers to wrangle about definitions, we can move to the questions that more practical scientists ask. Have paranormal events been seen, even at some low level, in carefully controlled, replicable experiments? What are the mechanisms by which supernatural or paranormal events happen, if they happen? In other words, is there evidence?

My Answer is No Surprise

By now my answer should be no surprise. No scientifically viable evidence has yet been found for events that violate known principles of physics or other natural sciences. Reviewing claims of such evidence has occupied the greatest portion of this book. By no means have I exhausted all the examples, but I think the ones chosen are representative and sufficient. They are the ones we hear most about. I doubt that other examples exist for which the verdict would not be the same. In any case, I feel no obligation to dredge them up.

Paranormal claims occupy far more space on bookstore and library shelves than their critiques, and the examples I have presented are the best that proponents of the paranormal have been able to come up with. It's a dismal record indeed. I am astonished that so many people in a modern nation like the United States still take the paranormal seriously. I shudder at what this fact implies about the general state of scientific education in America.

Occam's Razor

I have tried to make clear the careful process by which the methodology of science is applied to claims of new phenomena of any kind, normal or paranormal, for this is a great source of misunderstanding. The key methodological principle, which is repeatedly ignored by those who assert that various observations support paranormal interpretations, is Occam's razor: a new phenomenon must be required by the evidence, not just be consistent with it. Mere consistency implies that we can do just as well without the hypothesis. So why introduce it?

Simply put, no need has been demonstrated for a paranormal element to help explain any events so far observed. The most economical hypothesis, confirmed by every observation, is a universe of discrete bits of matter interacting locally by natural forces—and nothing more.

Pseudoscientists see evidence for their claims in the unexplained, the mysterious, and the anecdotal. Most scientists see only the unexplained, and insist that all plausible natural explanations first be exhausted before paranormal ones are considered. If a phenomenon remains unexplained for lack of data, calling it evidence for the paranormal is the height of irrationality.

The value of Occam's razor as a practical methodological rule has been confirmed throughout the history of science, where its application has kept people from accepting numerous claims that eventually faded away. I have discussed just a few of these, such as N-rays and cold fusion. I could list many more from conventional as well as unconventional science.

In the examples of psi phenomena discussed in this book, evidence for paranormal effects was widely acclaimed and then later found to have a more likely mundane explanation. The explanation in many cases turned out to be fraud, either proven or not sufficiently ruled out.

In observations where fraud or other nonparanormal explanations cannot be proved for lack of data, it is a serious mistake to take the observations as supportive of paranormal claims. This error is most common in the popular literature, though one sees it in supposedly scientific articles and books as well.

Many parapsychologists display their own true pseudoscientific colors when they encourage, and indeed profit from, this gross misuse of both science and the media. I cannot emphasize too strongly and too often that the burden of proof rests not on those doubting the claims of paranormal effects but on those proposing them. Scientists as well as pseudoscientists are guilty of fraud or incompetence unless they prove themselves innocent.

Incompetence

Fraud and incompetence have each played a disturbingly prominent role in psychic experimentation. Many experiments do not sufficiently rule out plausible normal explanations or fraud, when even the simplest and most obvious precautions might greatly reduce these possibilities. And testimonials from high places to the skill and integrity of an investigator, or other appeals to authority, do not take the place of hard evidence.

Examples of inadequately controlled research were seen in the experiments of Rhine and his collaborators in previous decades, and in those of Tart, Targ, Puthoff, Schmidt, and Jahn in recent years. As an experimental scientist, I am appalled by the poor design and execution of experiments that are proclaimed by the parapsychologists as the best work done in their field. How could Rhine not have had Pearce watched? How could Tart not have considered repeated digits? How could Targ and Puthoff have allowed so many clues to creep into their remote viewing tapes and not have kept an eye on Geller's confederates?

I don't think any of these people were stupid. I can only conclude that, consciously or unconsciously, they wanted to be fooled. The "will to believe" probably played an important part, as it did with the nineteenth-century psychic researchers who allowed themselves to be taken in by the sleight-of-hand tricks of spiritualist mediums. And unfortunately, money is to be made in books on the paranormal, palmed off on an unsuspecting public that trusts the printed word far more than it should.

Coming Up Empty

After a century and a half, the scientific search for the paranormal has come up empty. Unlike the numerous anecdotal stories that fill the pages of popular literature, the experiments conducted in laboratories over those years had the virtue of a modicum of controlled and systematic procedure. This has made it possible, in many cases, to find sufficient data to refute them. These were the good experiments.

The best experiments, like the best theories, are those that provide the rest of us with the ammunition to destroy them. If they are good, they will withstand the fire. The worst experiments, like Pearce-Pratt and Targ-Puthoff, hide the details, forcing critics like Hansel or Marks and Kammann to go in and dig the facts out for themselves. While no experiment in science is ever perfect in providing a complete account of itself, many in parapsychology have been particularly notable for failing to report adequate information to allow evaluation and duplication.

The Failure to Replicate

Ultimately, replication of experimental results by independent researchers must decide the validity of new phenomena. In respect to replication, psychic research has failed miserably. Rhine replicated Rhine, but that was not an independent replication. Soal claimed to have replicated Rhine, but was shown almost certainly to have altered the data. No one replicated Tart or Targ and Puthoff except themselves, and great doubt has been cast on their experimental methods. Jahn claims to have replicated Schmidt, but they disagreed on the quantitative level of the effect and also used flawed experimental protocols. We have seen similar stories repeated again and again throughout the history of parapsychology.

Little Impact

The failure of psi phenomena to replicate under controlled laboratory conditions is the strongest argument against its existence. But so far this argument has had little impact on the widespread public belief in the existence of psychic phenomena. This belief has no rational, scientific basis.

Rather, belief in psychic powers results from traditional teachings that go back to primitive times. Human beings have always, almost universally, believed in the existence of a world beyond the senses. As further support for this ancient belief, people attribute certain personal experiences to the supernatural, simply because they either fail to be aware of alternative natural explanations, or hold to invalid interpretations of nature. Or perhaps evolution has built these beliefs into our genes, by virtue of earlier survival value, making it very hard for us now not to believe—even though these remnants of a different time now endanger the survival of our species.

Many people tend to classify experiences as psychic or paranormal if they are strange, unfamiliar, or involve mental processes they wrongly assume to be nonphysical. Simple coincidence is most often the explanation, though people sometimes find this explanation hard to accept. But unlikely events do occur by coincidence far more frequently than people realize. In many cases, a coincidence probability can be estimated numerically.

If the odds are one in a million for an event to happen in a given time period, then this event will occur, on average, once in that period to some resident of a medium-sized city. Newspapers must learn to disregard reports of probably coincidental events, just as they disregard many other preposterous stories they receive each day.

Mind and the Brain

Beyond those natural events that people take to be paranormal because of their strangeness or rarity, mental phenomena are commonly assumed to have qualities that extend beyond the realm of the physical. Because we have taken so long to even begin to understand the operation of the nervous system, the ancient belief persists that mind exists as some kind of disembodied, spiritual entity.

But we have seen no evidence for a mind independent of matter. All mental processes, from emotions to creativity and intellectual thought, can be correlated with physical processes in the nervous system. These are affected by drugs, aging, and disease, and have been localized to different portions of the brain.

What Makes us Human?

Many people object to the notion that the human brain is a computer. Sure, they say, this may be true for functions such as logic or word processing, but humans are more than machines. They have emotions. Computers supposedly do not exhibit emotions.

Can we single out emotions as a quality that separates humans from machines? Hardly. In the first place, emotions are not a unique human quality. Animals have emotions. Studies of the brain demonstrate that our emotions are processed deep in the more primitive recesses of the brain that we share with other animals. Emotions such as anger and fright represent automatic responses to threatening situations, as blood is pumped to crucial areas of the body to allow for quick action. The survival of the species is made possible though these emotions, as well as others such as love and lust.

Even before the days of the computer, automatic machines were, in a sense, emotional—designed to respond rapidly in a particular way to a specific stimulus. Far from making us uniquely human, our emotions bind us even more closely to our animal (and mechanical) brethren.

What makes us uniquely human is not emotion, but our remarkable intellect. The magnificent serial computer in our left cerebral cortex, and the equally majestic parallel processor to its right, operating in concert with the rest of the nervous system, provide us with the capability to gain control over our baser instincts. The great power of the human nervous system is its ability to react in a variety of ways, optimizing responses to complex sets of stimuli, enabling us to choose among possible actions in a responsible, thoughtful way. This is the source of our freedom of will.

The Dehumanizing Effect of Supernatural Belief

The materialist view is often attacked as dehumanizing. Rather, the opposite is true. Supernaturalism, religious or secular, dehumanizes us by making us automatons. In the supernatural view, humans are controlled by forces outside themselves. Events happen because they are predestined.

Those who believe that certain people have the power to see into the future make the assumption that the future has been determined. They dehumanize us because they maintain that we have no choice in what happens. They believe that whatever will be, will be. The earthquake destroyed the city because "it was written." The bullet that killed the soldier in battle "had his name on it." If you walk across the street and are hit by a car, it was "fated" to happen.

Materialism, on the other hand, humanizes each of us. It restores our dignity as individuals with a certain measure of control over our own destinies. The future is not determined. What will happen to each of us tomorrow depends to some extent at least on what we chose to do today. You can prevent that car from hitting you by pausing to look before you cross. You can help stave off cancer and heart disease by quitting smoking.

We each lead better lives by working to make them better. We are not under the controls of any supernatural forces. While bound by the limitations of physical law, we still possess considerable free choice. I don't know about you, but· I find this view infinitely preferable to supernatural determinism.

For those religious readers who pity me for my lack of faith and offer to pray for me, please don't. I appreciate your concern, but it is quite misplaced. I will be very disappointed if I someday discover that I was all along controlled by outside forces, that the daily choices I have made throughout life were not really my own. I believe I am really free. Materialism makes me human. Materialism makes me free.

Mystical Experiences and Revelation

The so-called mystical or spiritual experiences that are traditionally attributed to supernatural human capabilities correlate, like all other mental phenomena, with brain functions and dysfunctions. For example, many reported religious visions over the ages have all the earmarks of epileptic attacks or hysteria, or may have been drug-induced. Furthermore, no purported insight obtained during meditation, trance states, or seizures has ever been demonstrated to contain information that could not have existed all along within the brain of the individual.

I have given examples of fundamental knowledge about the universe that were never even hinted at in the claimed revelations of the world's greatest spiritual leaders. Not only has revelation proved unproductive as a source of knowledge about the universe, it has served to impede progress in knowledge, as an unholy alliance of churches and kings forced erroneous views of the universe on humankind for millennia. Only when people were able to gather the courage and resources to fight back against ignorance and superstition did the rapid accumulation of knowledge we associate with the scientific revolution take place.

Still, the belief that knowledge can somehow be obtained by paranormal means persists. I have issued a challenge to New Age channelers or any others claiming that they are in contact with spirits or superhuman entities, to prove their claims by asking their sources to provide the answers to certain outstanding fundamental questions about the universe. All of these questions should be answered by the normal development of science within the next decade, so this test can provide us with definitive proof of a world beyond.

A Theoretical Vacuum

In particle physics, as in most sciences with which I am even vaguely familiar, the experimental search for anomalous phenomena does not proceed in a theoretical vacuum. Theory, after all, defines what is anomalous. It guides us where to seek the information needed to proceed to the next level of understanding. And so, for psychic phenomena to be confirmed, this development will have to be accompanied by the parallel development of a viable theoretical framework.

Suppose we make an observation that violates established physical principles. The result, even if truly anomalous, can still be natural and fit into a new set of physical principles yet to be worked out. Indeed, as we have seen, anomalies often point the direction in the search for new physical principles.

Many psychical researchers over the years have left open the possibility that psi phenomena will turn out to be consistent with suitably modified physical law. Thus, we have seen how Oliver Lodge thought that psychic fields were connected to the aether, the continuous fluid that he and his contemporaries believed pervaded all of space. But Einstein put this hope to rest with the special theory of relativity and the photon theory of light.

Mathematical Tools

As I have noted, the classical gravitational and electromagnetic fields of nineteenth-century physics have become the mathematical tools for describing particle interactions in the twentieth. They have no reality other than mathematical, though in physics classes they are normally presented as real entities—contributing greatly to the confusion that is exploited by paranormalists. In no experiment do we directly observe gravitational, electric, or magnetic fields. Rather we observe localized bodies that move as the result of actions of other observed localized bodies. These bodies might be real or they might be metaphors, but the field vectors of classical physics are certainly metaphors.

In the twentieth century, classical electromagnetic fields have evolved into the abstract wavefunctions and even more abstract state vectors of quantum mechanics, which have no more external reality than the symbols used to represent them in a quantum mechanics textbook.

The action at a distance that was formerly described by means of classical fields is now understood to consist of particle exchanges between interacting particles. When electron A repels electron B, it does so by emitting a photon that is absorbed by electron B, causing electron B to recoil backward under the impact. The exchanged particles cannot travel faster than light, and so, instantaneous interaction or communication between different points in space is impossible.

We have seen how, in recent years, paranormalists have made the same error Oliver Lodge made in the past. They still, at least implicitly, assume that the universe is pervaded by some kind of continuous fluid.

In Kirlian photography and other techniques for displaying what are claimed to be the auras surrounding human beings and other living things, we saw that only simple, perfectly understood, electromagnetic effects are involved. Only complete ignorance of even high school level physics permits anyone to take these so-called auras seriously. A theoretical basis for cosmic consciousness certainly does not lie in classical electromagnetism, any more than a theoretical basis for astrology can be found in Newton's 300-year-old law of gravity.

The Quantum Cosmic Fluid

A considerably more sophisticated theoretical framework for psi phenomena can be found in the ideas surrounding quantum mechanics, but we have seen that this, too, is the result of a lack of understanding of the subject. Some hope for the existence of an unseen cosmic fluid lay in the hidden variables ideas of David Bohm and his followers. But Bell's theorem provided a definitive empirical test.

The experiments of Aspect and others that tested Bell's theorem failed to confirm the existence of hidden variables, obtaining results totally consistent with what has been, for over fifty years now, the prevailing point of view of the overwhelming majority of physicists.

As I have noted, this is not the impression that one gets in reading paranormal literature. Numerous authors, including a few physicists who should know better, would have us think that the confirmation of the spooky action at a distance of conventional quantum mechanics has opened a door for psi phenomena. I believe I voice the consensus of the majority when I claim the contrary: the experiments have slammed shut this particular door.

The experimental tests of Bell's theorem gave results in precise quantitative agreement with the predictions of the Copenhagen Interpretation of quantum mechanics. The experiments rejected the most important class of hidden variables theories, namely those in which the variables are local and real. Nonlocal or nonrealistic hidden variables might still exist, though we have no evidence that they do. Perhaps paranormalists can still make something of that, sneaking psi through the crack under the closed door. But I would wager that anything that filters through will be experimentally indistinguishable from conventional quantum theory.

Voodoo Science

The empirical confirmation of the conventional interpretation of the wavefunction as a nonreal mathematical entity has not left much theoretical room for psychic phenomena, despite the grand claims. I have discussed the brave attempt to connect consciousness to the collapse of the wavefunction, and thereby provide a mechanism for PK—mind over matter. This attempt also fails at every level.

To go from the statement that physical quantities have no meaning until they are measured, to the assertion that human consciousness thereby affects these quantities implies a gross misunderstanding of quantum mechanics. First, the measurement could be done by a machine just as well as a person. Second, physical quantities are not intrinsically real. They are determined by the measurement process itself. Time is what you measure on a clock. How we design the clock determines what we measure. The same is true for spatial position, mass, energy, and all the other quantities we have invented to describe our observations. For example, many different temperatures exist (thermodynamic, bolometric, kinetic), depending on how the measurement is made.

Now the reader might say that I have returned to Cartesian idealism with all these unreal mathematical quantities: gravitational and electromagnetic fields, wavefunctions, space, and time. Am I not saying that the universe is all in our heads? No, I am not. I believe that an external reality exists out there and will still be there long after the human race is extinct.

But the only way we still-existing human beings know anything about that reality is by means of the data this real universe presents to our five senses.

Physicists don't claim to associate external reality in strict one-to-one correspondence with the concepts that we have invented to describe sensory data. These concepts are in our heads, and are abstract representations of the reality.

But I am really saying the obvious, I think. Is the name of a person equal to the person? Is his or her image on film, or in the form of a voo-doo doll, deeply connected to that person, so that sticking a pin through it will cause harm? The idea that we can control the behavior of the mate-rial world by merely thinking about it is voodoo pseudoscience. But then, parapsychologists do voodoo, too, do they not?

A Body in Kahuku

While writing this chapter, I received a call from a wire service reporter. He said he wanted to alert me to a story that was breaking in Honolulu. A young woman from Kahuku, the main sugar-producing area left on Oahu, had been missing for several days. The early news that day was that a search organized by the family had found a body just off a road in heavy brush. The body was later confirmed to be that of the missing woman. The reporter said that a local psychic claimed to have led the family directly to the body, and that this claim was confirmed by the family. The family also said that the psychic had communicated with the woman's "soul." The police would only say that the psychic's description of where to search could have been anywhere on the island.

I thanked the reporter for the information and told him that I was not about to tell a grieving family what to believe. Obviously, they found great comfort in the psychic's story, as did William Crookes, Oliver Lodge, and many others throughout history when confronted with the death of a loved one. It is perfectly understandable to want to believe that a loved one still lives and that she will someday be united with her family again.

But the story made me angry at the psychic who was using the tragedy of a family for her own personal gain. Her prediction only was that the body would be found in high grass. The family had been searching fruit-lessly for days in places suggested by the psychic, and had previously spent most of one day searching a valley far from the site where the body was eventually found. Certainly the prediction that the body was in high grass was not very risky. The woman was from Kahuku, after all, and cane fields are "high grass." If she had been in the open or low grass, she would have been found a lot sooner. A successful prediction that the body was at a specific address, or on top of a particular mountain, would have been much more impressive.

My anger at the psychic extended to all the psychics, astrologers, New Age channelers, faith-healers, TV evangelists, mind expanders, and others who exploit the sick, bereaved, or simply unhappy.

I must add that my anger does not extend to the many legitimate clergymen, nuns, and other honest participants in the religious enterprise who have demonstrated the sincerity of their faith by examples of their personal sacrifice and commitment. Selfless figures such as Mother Theresa and Father Damien deserve our highest respect.

I can understand how, because of the traditional teachings of the world's cultures, many become convinced that a better world must exist, and that they can do nothing more profitable in this world than help others to achieve the next one. But those few others on the fringes, who exploit the deepest needs of people for their own personal enrichment, rank in my mind with criminals as enemies of civilized society.

God Will Provide

I am a scientist, and although other scientists claim to be able to separate their science from their religion, or to reconcile the two, I cannot. I find it impossible to accept on faith what strikes my reason as being too pre- posterous to believe. Without a single shred of evidence that is not compromised by one of the many flaws we have noted in this book, and seeing no intellectual basis for such belief, I cannot believe on faith alone.

So what can I offer to the ill and the bereaved, or the healthy and happy who know that their health and happiness are fleeting? First, I would offer the opinion that the answer cannot lie in faith in a probably nonexis- tent superworld. What possible good could come from living a fantasy?

Placing their faith in the supernatural, people often will not take the natural measures necessary for their own well-being. I am reminded of a student with crucifix earrings who once visited my office to get my signature on a course drop slip. He smiled beatifically, saying he guessed he was just stupid and that's the way it was. Perhaps if he didn't believe that Jesus had made him stupid—for the Lord's own mysterious reasons—the student might have made a little more effort and passed the course.

I am also reminded of a story I saw years ago on the TV program "60 Minutes." An elderly southern black lady had lost her house for failure to pay a $20 tax bill. The house was put up for auction and was bought by speculators for a fraction of its value. The woman was very religious. She spent most of her day in prayer. When asked why she didn't pay the bill, her answer was, "God will provide." Well, God didn't.

American culture today rests on fantasy. We are exhorted by advertising to buy every conceivable type of item on the expectation that these will make us as beautiful or successful as the perfect people in the ads. In movies we are treated to a continual parade of stories about people who win out over great odds, who have their dreams come true. Notice how often these films have supernatural themes, and use spirits or miraculous events. I enjoy some of them myself, but I wish they were balanced with a little more reality (although Ingmar Bergman is a little too much reality for my tastes).

We are exhorted by authors to believe that everything is possible, that we can all be superstars, get rich, or become president, if only we try hard enough. Well the world isn't this way, and the sooner we realize it, the better. We aren't Peter Pans who just have to clap our hands and our wishes will come true. Most of us are not beautiful or talented. Most will not get rich or marry a prince. Life is hard enough, without having to live it with the additional burden of wondering if we are at fault for not being the great success that advertisers tell us we can be.

Vessels of Information

We cannot live in fantasy. We must accept our individual mortality, our finiteness, our structure of localized atoms. Humanity was an accident that happened as the product of a series of random events in a universe so vast that even the most unlikely event can occur without cause or plan. We must learn to live within the framework of this understanding. It's not that hard. In fact, it's much easier than you might think.

Perhaps the greatest step we can make in the direction of acceptance of our ultimate fate is to recognize and understand our true role as individual organisms in the overall scheme of things. Our bodies are vessels. Not vessels for a supernatural soul, but vessels that carry the genetic information of billions of years of life on this planet. Perhaps that information constitutes the closest thing we have to an immortal soul. But if so, this soul is not unique to the human race. It is common to all life.

Within most cells of every complex organism on earth is the history of our tribe, not just the human tribe but the tribe of all living things. This record is written in our DNA, and is transmitted from generation to generation over a time span that is almost impossible to imagine. For example, the DNA text of the histone H4 gene, 306 characters long, goes back over a billion years to the time of the common ancestors of both animals and plants.

This consciousness of our role as data carriers is far different from the traditional teachings of the world's great religions, with the possible exception of the original form of Buddhism, stripped of the supernatural mumbo-jumbo added over the centuries. Recognizing who and what we really are should lift a great burden from us as individuals. We do not have to all be saints, or geniuses, or presidents, nor feel guilty or ashamed because we are not.

We simply have to be faithful carriers of our genetic information, seeing to it that the cycle continues by providing for the future generations who will carry it forward. The cycle will likely continue as it has for billions of years, perhaps leading us one day in the far distant future to the Omega Point of Teilhard de Chardin, provided we humans do not first commit the ultimate folly and destroy all life on this planet. Unfortunately, this is within our power.

The existence of this wonderful accident of nature we call life is now direly threatened. Because the human species has become so powerful, it is expanding to the point where grave changes to the planetary environment are taking place as the direct result of our overpopulation and other excesses. Because of our voracious appetite for energy, we are destroying the forests that keep the atmosphere in balance, pumping pollutants into that atmosphere and the oceans, and destroying many forms of life.

Why are we so shortsighted? At least part of the blame must be placed at the feet of religion. The Judaeo-Christian belief that humans are creatures made in the image of God has helped us justify our rape of the planet. In conjunction with this belief, it was taught that the earth was created for our use and we were given dominion over all of it. This allows us to take from the earth and its lifeforms as it strikes our fancy. It also gives us a highly inflated view of our individual importance.

Judaeo-Christian ethics did little harm when we were few in number; in fact, it was itself probably the product of the natural genetic imperative. We have seen that religion may have originally developed as a mechanism to aid in the survival of human genes. Wouldn't that be an irony? Religion a product of evolution!

However, beliefs that were of value in a primitive society are dangerous baggage in the modern world of today. For example, today's Catholic and fundamentalist Christians contribute to overpopulation by strongly resisting birth control measures that, if practiced more widely, could greatly slow the population explosion. They justify their position on this matter in the name of the "sacredness of life." But how sacred will life be when it no longer exists?

Removing the Yoke

Thankfully, evolution has prepared us to take the actions necessary to guarantee the survival of life, if only we can overcome the forces of superstitious belief that currently act against our best interests, leading us down the path of destruction. Our genes are programmed so that the survival of our genetic information has higher priority for us than our individual selves. So it is natural for us to sacrifice our lives for our children, other loved ones, our tribe, or some great cause. To do this without the promise of reward is a truer altruism than living a life of sacrifice for others in the hope of ultimate eternal bliss.

When we ask for individual immortality, like Faust, we sell our soul to the devil. The soul we sell is not some disembodied spirit, but our freedom to live and act according to the fullest knowledge about the world and our own best instincts. The supernatural has been a yoke on the neck of humanity since we first began to think and dream. Now those same thoughts and dreams are pointing the way to remove that yoke.

Index